Knowledge Management in the Innovation Process

Economics of Science, Technology and Innovation
VOLUME 24

The titles published in this series are listed at the end of this volume.
Economics of Science, Technology and Innovation

Knowledge Management in the Innovation Process

Edited by

JOHN DE LA MOTHE
PRIME, University of Ottawa

And

DOMINIQUE FORAY
CERI, OECD

KLUWER ACADEMIC PUBLISHERS
Boston / Dordrecht / London

Distributors for North, Central and South America:
Kluwer Academic Publishers
101 Philip Drive
Assinippi Park
Norwell, Massachusetts 02061 USA
Telephone (781) 871-6600
Fax (781) 871-6528
E-Mail <kluwer@wkap.com>

Distributors for all other countries:
Kluwer Academic Publishers Group
Distribution Centre
Post Office Box 322
3300 AH Dordrecht, THE NETHERLANDS
Telephone 31 78 6392 392
Fax 31 78 6546 474
E-Mail <services@wkap.nl>

Electronic Services <http://www.wkap.nl>

Library of Congress Cataloging-in-Publication Data

Knowledge management in the innovation process/ edited by John de la Mothe and Dominique Foray.
p. cm. -- (Economics of science, technology, and innovation ; 24)
Includes bibliographical references and index.
ISBN: 0-7923-7464-9 (alk. paper)
1. Knowledge management. 2. Technological innovations. I. de la Mothe, John. II. Foray, Dominique. III. Series.

HD30.2 .K6373 2001
658.4'038--dc21 01-038305

Printed on acid-free paper.

Printed in the United States of America

CONTENTS

CONTRIBUTORS

Nabil Amara, Département de science politique, Université Laval.

Michael Bordt, SIEID, Statistics Canada.

John de la Mothe, PRIME, Faculty of Administration, University of Ottawa.

Meinolf Dierkes, Wissenschaftszentrum Berlin für Sozialforschung, (Social Science Research Center, Berlin)

Dominique Foray, CERI, OECD, Paris.

Réjean Landry, CHSRF/CIHR Chair on Dissemination and Uptake of Research, Département de science politique, Université Laval.

Stéphane Lhulliery, C.E.P.N., C.N.R.S., Paris

Matthew Lucas, PROGRIS, Centre for International Studies, University of Toronto.

Larry Prusak, IBM Institute for Knowledge Management, Boston.

Hans G. Schuetze, University of British Columbia.

Nico Stehr, Kulturwissenschaftliches Institut, Essen.

Kathy Wallace, Canada Institute for Scientific and Technical Information (CISTI), National Research Council of Canada (NRC).

David Wolfe, PROGRIS, Centre for International Studies, University of Toronto.

ACKNOWLEDGEMENTS

This book is the fourth in a series of joint research projects, organized by Program of Research on Innovation, Management and the Economy (PRIME) at the University of Ottawa and Statistics Canada, which have examined key aspects of the innovation process. The intent of the work is not only to improve, in a scholarly sense, our understanding of the social and economic dynamics of innovation but is also to raise important questions as they pertain to measurement, research and policy issues. Each volume has emerged out of a Research Workshop which brought together practitioners from government and industry along with international experts in the field. This book is no exception. The first volume - *Local and Regional Systems of Innovation* (1998) - studied locational issues related to innovation. The second and third - *Information, Innovation and Impacts* (1999) and *The Economic and Social Dynamics of Biotechnology* (2000) looked at two different transformational technologies. This volume begins to look inside the 'black box' (to borrow Nathan Rosenberg's phrase) of the knowledge economy, namely to look - as the title suggests - at Knowledge Management in the Innovative Process.

The organization and preparation of this volume gave rise to a most pleasurable occasion for the two editors to work together for the first time. Each had known of the other's work for some time but 'knowledge management' seemed to give us an excellent 'excuse' to prove how the 'networked economy' can actually work between Ottawa and Paris.

As with the previous work, turning the workshop into a book required the involvement of a number of talented and generous people. The workshop, which was held in Ottawa in February 2001, gave the contributors featured herein a rich opportunity to interact and exchange ideas. To all participants, observers, and translators, thanks must be given. Essential logistical and research support was provided by Michael de Sa, Ana Ranert, and Justin Taylor of PRIME. Tyler Chamberlin, a doctoral student at the University of Manchester (CRIC) and a Research Fellow at PRIME once again expertly prepared the manuscript for publication. Tom Randall and Marian Scott of Kluwer gently guided the book throughout the publishing process. And Bo Carlsson and Cristiano Antonelli have once again invited us into their fine series. Thanks to all.

JdlM/DF

PREFACE

This is the fourth book from a project that examines technological change from economic, social and statistical perspectives. The previous three books dealt with innovation, the first with its regional aspects, the second and third with the impacts of two transforming technologies, the information and communication technologies (ICTs) and biotechnology. The present volume continues this project with a focus on innovation resulting from the use of a set of practices, rather than a technology. The practices are those used to manage knowledge.

Knowledge, in some form, has always been managed. However, the pervasive ICTs permit firms to gather information about their suppliers, their clients, and about the way in which they add value to what comes from the suppliers before the product goes to the client. This is a basic command and control model but it becomes more complex if the firms use networks to help them navigate through the social, economic and regulatory environment. The knowledge about where the firm is going and how it plans to get there is facilitated by ICTs, but that is not all there is to the management of knowledge. Knowledge management involves practices such as space for sharing, incentives for sharing, and different mechanisms for sharing such as story telling.

In a global economy, the study of the practices of knowledge management cannot be done in isolation if there are to be meaningful comparisons across industrialized countries. For some years the OECD has been conducting forums on knowledge management in public and private institutions and, following the forum in Ottawa in September 2000, a decision was taken to develop a survey of the use knowledge management practices. The first meeting to develop a questionnaire was held in Copenhagen early in February 2001 and it was followed by a second meeting the day before the research workshop which gave rise to this book. The next steps are consensus on the questions to be put to firms in OECD economies and the mounting of pilot surveys in selected member countries to learn how to ask the questions and interpret the answers. This book is a contribution to that process.

Knowledge management practices are used by firms to do better what they do, to innovate, to become more productive, more flexible, and more socially responsible. Understanding these practices is an important goal and it was for this reason that a research workshop was convened jointly by Statistics Canada and the Program of Research in Innovation Management and Economy (PRIME), of the University of Ottawa. The papers in this book resulted from presentations at that workshop.

The Introduction relates knowledge management to the process of innovation and provides a context for the book. It leads into the framework section and the two papers which provide ways of looking at the insights from the papers in the other sections. The first provides a systems model of the firm interacting with its environment at three levels: a basic response to market signals; a tactical response requiring a basic analytical function; and, a strategic response requiring vision and foresight. The second paper reviews the reasons why former methods of knowledge management are no longer working and why there are strong incentives to adopt new methods for the new economy.

The section on measurement presents three aspects of knowledge and measurement. The first deals with the flow on knowledge through intellectual property mechanisms, the second introduces the important concept of social capital and its role in innovation and the third provides an overview of French experience of measuring knowledge in the 1990s.

While most of the presentations at the workshop dealt with the understanding of knowledge management, Larry Prusak brought to the meeting years of experience of promoting the practice of knowledge management. His paper provides insight into what works, and what does not, and complements the papers on measurement and analysis.

The section on impacts goes on to look at how knowledge management is used and how knowledge is transferred in different contexts, this is followed by a reflection on productive knowledge and the section ends with some policy implications of promotion and use of knowledge management practices by firms.

The workshop was fortunate to have Keith Pavitt of the Science Policy Research Unit of the University of Sussex both as a participant and as the provider of a synthesis of the ideas that were raised in the course of the meeting. Many of these thoughts made their way into the concluding chapter which goes on to suggest research directions for statisticians, for policy analysts, and for academics.

Fred Gault

Director, Science, Innovation and Electronic Information Division
Statistics Canada

PART I

INTRODUCTION

Chapter One
APPROACHING THE MANAGEMENT OF KNOWLEDGE

John de la Mothe
PRIME, Faculty of Administration, University of Ottawa

and

Dominique Foray
CERI, OECD

1. INTRODUCTION

Many of the unresolved problems in economic policy, management and research turn on questions of knowledge. This is not new, but it becomes more important as technologies facilitate the transmission and the use of knowledge and make its management a practical reality. This raises questions, which this book addresses. However, it is not a book about technology. It is about knowledge, its management, and the implications of knowledge management for innovation.

Knowledge can be seen as lurking decidedly behind Adam Smith's (1776) theory of the division of labor (Porta, Scazzieri and Skinner, 2001). Since then, much of the philosophy of science debate - for example the famed Popper-Kuhn debate (Lakatos and Musgrave, 1970; Rugman, 1998) - has expressly dealt with the growth of knowledge.

This theme, the production of knowledge, suggests that there are shifts in the ways in which scientific, social and cultural knowledge is produced and used. This in turn suggests a shift in the nature of institutions, disciplines, practices and policies (these are all themes which resonate throughout this volume). This 'shift' is now emphasizing transdisciplinarity, complexity, heterogeneity, connections and the diffusion/use or application dynamic of knowledge. As a result of this analysis, the changing nature of knowledge is resulting in qualitatively new social relations and social contracts between social institutions (including firms), economic performance and activity, and technological design and trajectories.

A derivative impact of this sort of interest can be seen when popular pundits such as Robert Reich - formerly Secretary of Labor under U.S. President Clinton and now professor at Brandeis University - and Californian management statesman Peter Drucker proclaim that knowledge is the asset to which we must attend. As Reich has put it, (1992) the currency of today's social and economic analysis is learning. Put similarly, in *The Post-Capitalist Society* (1993), Drucker has stated that "knowledge is the only meaningful resource today".

More in the day-to-day observations of attentive news watchers and economic analysts alike, knowledge has become the central factor of production in the advanced industrial nations. This can be anecdotally seen in the widespread de-materialization of trade, in the shift in job creation towards knowledge-intensive services, in the emergence of e-commerce, and in the internationalization of knowledge production. Not only is this rapidly developing environment demanding a re-calibration of neo-Ricardian trade theory, ideas of value added and of productivity, but it is also dramatically challenging ideas and practices of management.

Practical and direct responses to this can be seen in business with the rise of the Chief Knowledge Officer (CKO) or the Chief Information Officer (CIO) as a new position within complex and multinational firms across sectors and functions. It is not always clear, however, what these individuals are expected to do or achieve. A not-unrelated trend can be seen in changes in government practices, for example in the move towards Alternative Service Delivery (ASD), horizontal coordination, data sharing and function alignment.

Both types of institution face similar problems. They must ask, and answer, such elusive questions as 'what is knowledge?'; 'where is knowledge?'; 'who has the knowledge?'; 'does the organization lose or gain competitive advantage or effectiveness by sharing or leaking knowledge?'; 'where can we find the knowledge we need?'; and so on. Scholars often highlight the differences between knowledge, information and data. To this we can add the critical question as well, oft' asked by statisticians and measurement officials, 'how can we measure knowledge?'. In a knowledge-based economy, no policymaker, analyst or strategist can easily afford to avoid such queries.

To those with an evolutionary bent, it is familiar now to approach such important issues by way of differentiation. For example, it is common - though not easy - to differentiate between codified (or explicit) knowledge and tacit knowledge. It is prevalent - particularly in literatures associated with innovation systems, clusters and technopoles or technological systems - to see 'ladders' or taxonomies which differentiate between 'know how', 'know what', 'know who', 'know why', and so on. Analytically, and from a measurement perspective, these are valuable and useful types of distinction.

Nonetheless, a number of literatures and communities are stymied by the complexity and dynamics of the problem. In much of the cybernetic, chaos and systems-theoretic literatures for example, knowledge and knowledge management are seen as simple questions of engineering and as being open to modeling (Schreiber et. al., 2001). Such approaches do find valuable application in inventory and production management, but they tend to narrowly limit what complex organizations' interests are in managing knowledge or reduce them to algorithms.

Network theorists (cf. Beckmann, Johansson, Snickars and Thord, 1998) are more accepting of messiness. This community is quite articulate in discussing chance, noise, co-adaptation in networks, as well as the ascendance and decay (shift) of leadership in organizations, for example. But the proximity of interest in or applicability too management, strategy, measurement or policy is remote.

To those scholars emerging from the legalistic, strategic studies or international relations literatures, questions of knowledge management often get 'shoe-horned' into questions of 'knowledge diplomacy', but these rarely meet the needs of knowledge managers (Ryan, 1998). Similarly, to many social scientists knowledge management typically conjures up frameworks which are shaped (captured) by hegemons, vested interests, power and control. But such 'guiding' diviners are pre-destined to reinforce the frameworks à la Gramsci (Gill, 1997) instead of guiding the managers, policymaker and statistician.

2. KNOWLEDGE, MANAGEMENT AND INNOVATION

In the firm, one of the reasons for managing knowledge is to accelerate the introduction of new or significantly improved products or processes as part of its commercial activity. This is innovation (OECD, 1996b).

The following chapter by Dierkes provides a classification of firm activities which also supports the analysis of innovation. The classification starts with a firm in a stable environment. The next level is instability in the local environment which still permits the firm to maintain its vision unchanged and still function. The third level addresses the global environment of the firm and the need to benefit from both hindsight and foresight and to have the flexibility to adjust its vision and purpose in light of broader socio-economic interests.

From the innovation perspective, a firm that exists in a relatively stable environment will manage the knowledge that it gathers about prices and quantities of its products in a competitive market, and of the intermediate goods and services purchased from suppliers. It will also manage its knowledge of the production process and it will continously improve its processes and products. In this type of firm there is unlikely to be a need for radical innovation and the strategy and vision of the firm remains unchanged.

At the second level, the local environment is changing, but not enough to destroy the firm. However, to operate in the changing environment, the firm must be able to change its strategy and have the management capacity to implement that change. Part of this strategy can be the introduction of radically changed products or processes. The management of the knowledge of the firm expands from the first level to embrace the management of change, within the vision of the firm.

The third level adds time and space dimensions to address wider social and economic changes and requires the managerial capacity to manage the knowledge gained from crises in the past and to support foresight and informed choice. Here the firm must be prepared to reinvent itself and to change its vision. In such firms, new products and processes are anticipated which could have major socio-economic impacts. In these firms the management of knowledge becomes a central tool for operating and for decision making.

In all of this analysis, the focus is on the management of knowledge, not on the use of technolgy, however important that might be in support. Prusak in chapter eight reminds us that a project that spends more than a third of its budget on technology is not a knowledge project. It is a technology project.

3. CONCLUSION

Hence it is the interest and goal of this volume to seriously inspect the nature of knowledge management and to engage these notions with the processes on innovation. In concert with the authors cited, we hold that knowledge is the new coin of the realm. What we strive to achieve here, however, is to examine frameworks, measurement issues and impacts that are of direct import to the manager, strategist, analyst, statistician, and policymaker alike in an effort to better appreciate the dynamics of innovation and the knowledge economy.

PART II

FRAMEWORKS

Chapter Two

VISIONS, TECHNOLOGY, AND ORGANIZATIONAL KNOWLEDGE:

An Analysis of the Interplay Between Enabling Factors and Triggers of Knowledge Generation

Meinolf Dierkes
Wissenschaftszentrum Berlin für Sozialforschung
(Social Science Research Center Berlin)

1. KNOWLEDGE AND ORGANIZATIONS: WHAT IS NEW? WHAT IS RELEVANT?

Knowledge management, as the most recent offspring of the study of organizational learning, seems to have developed as one of the hot fads in academia and business. Many publications have appeared during the last few years (Dierkes et al. Second Edition, 2001), ranging from a quite abstract treatment of the role of knowledge in organizational learning and transformation to efforts to motivate businesses to pay more attention to the knowledge base of the organization and the continuous development of its intellectual resources. A wide range of devices to improve and enhance the use of the knowledge base of an organization have been recommended. They encompass such diverse concepts as the 'chief knowledge officer' (CKO) or the 'corporate university', which are mostly a new label for what used to be called the human resource and management development function. The common understanding in most of the literature is that the knowledge base of an organization is more important than ever before to ensure its survival in the competitive global market (Nonaka & Reinmöller, 1998).

Technology also plays an important role (Büchel & Raub, 2001). A significant number of publications argue that intranet and internet support of "the knowledge link" in companies enhance the availability of knowledge throughout the organization. The assumption is that a broad and widely accessible knowledge base significantly improves individual and organizational decision-making and is particularly important in a period of rapid and unpredictable change, as the current era commonly is described (Nonaka et al., 2001b).

Knowledge and the knowledge-based organization are clearly part of today's 'in speak' in management and among organizational development specialists and consultants. Despite all this faddishness, however, strong and convincing arguments in the literature suggest that knowledge - which has been an important determinant of competitiveness in a number of industries for a long time - is becoming increasingly central to larger segments of the economy. Technological challenges, globalization, and rapid socioeconomic and cultural change are seen as some of the major driving factors requiring a more solid knowledge base for the successful adjustment, growth, and survival of companies. However, the unpredictability of many of these changes also requires companies to increase their ability to integrate new knowledge and shed outdated knowledge bases (Dierkes, Hofmann, and Marz, 1998).

A strong organizational vision, a culture cultivating learning and sharing of a common knowledge base, a structure facilitating the wide use of individual and group knowledge, and leadership that fosters learning are seen as the determinants for creating a knowledge-base organization. They must be complemented by a general ability and willingness of individuals and groups to learn and by the companies' readiness to invest in individual and group learning and in the relevant technology base.

In this contribution I aim to explain one of the crucial, but under researched, fields in knowledge management: the links between visions and knowledge generation. Using selected concepts of past and potential overarching technological overarching visions and elements of organizational visions, I illustrate or, in the tradition of qualitative case-related research, test hypotheses from the organizational learning and knowledge creation literature. I conclude with a research agenda that should shed light on most of the key questions in this field which are still under researched.

2. KNOWLEDGE GENERATION, VISIONS, AND TECHNOLOGY: A CONCEPTUAL FRAMEWORK

2.1. The Background

From sustainable development and lean government to the learning society and the virtual company, visions as guiding concepts and the metaphors central to them have gained a great deal of attention in both the public at large and the business community (Dierkes, Hoffmann, and Marz, 1996). This growth in general interest has prompted a variety of academic disciplines either to begin to explore the concept of visions or to intensify research on the relevance, function, and explanatory power that visions have in social processes (Grin & Grunwald, 2000).

In management science and, to some extent, business history, the treatment of visions focuses predominately on their role and function within the organization. In this context visions are usually characterized as being closely linked to charismatic leaders or an organization's founding generation. According to House and Shamir (1992), "it can be safely concluded that there is a strong convergence of the findings from studies concerned with charismatic leadership and those concerned with transformational and visionary leadership" (p. 84). They suggested that leaders create visions to define the basic goals and strategic intents of their organizations. Since the mid-1990s, authors of management literature have credited visions with a role in the success of organizations in the increasing uncertainty of today's business environment. Scholars have recognized that organizations require not only strong leadership but also a clear vision to coordinate, direct, and guide them to achieve their goals. As Grossman and King (1993) stated, "visionary leadership is not a luxury but a necessity as we compete in the global marketplace with countries that routinely plan in terms of decades" (p. 56).

In both the business community and academic literature, this treatment of visions fails to address three important topics. First, it leaves certain factors in the environment of organizations unconsidered, especially broad and overarching visions in society, which influence organizational visions. Second, it seldom refers to the experience that visions often exist apart from the presence of charismatic leaders and also outlive founding generations. Third, it offers no satisfactory explanation of how and why visions shape the life and behavior of an organization.

These gaps should be filled by theorizing about how visions influence a key dimension of organizational life - the ability or inability of organizations to change perceptions, strategies, and behavior to meet the challenges of a rapidly changing environment. In short, how do visions affect the capacity of organizations to acquire and use new knowledge?

2.2. Organizational Learning and Knowledge Generation

The conceptual literature does not clearly distinguish between organizational learning, knowledge acquisition, and knowledge creation within or for organizations. Though differences could be elaborated upon, I will follow tradition and use these terms as they relate to the acquisition of information, the test of the relevance of the information, and the way in which information that is perceived to be relevant is converted into knowledge, made available in the organization, and used by organizational members. One can distinguish largely between three types of learning leading to knowledge generation in organizations (Pawlowsky et al., 2001). Though no unified nomenclature has yet emerged, most authors writing on the subject follow

Bateson's (1972) typology - Type I, Type II, and Type III - or at least their reasoning can be easily integrated into it. Probably the terms most widely used to characterize the three types of learning and knowledge acquisition are those introduced by Argyris and Schön (1978): single-loop learning, double-loop learning, and deutero-learning.

Bateson's Type I learning, or single-loop learning according to Argyris and Schön, is based on a process of error detection and correction aimed at improving or modifying the strategies and behaviors within the existing structure of the organization in a way that will best enable it to achieve its goals. Although some changes may occur, they do not alter the underlying norms of the organization. The organization continues to pursue its given policies and objectives. In this context, learning, and thus new knowledge, promotes a certain continuity.

Bateson's Type II learning, to which Argyris and Schön's double-loop learning corresponds, differs from Type I in that the organization recognizes environmental changes that cannot be responded to adequately within the existing context of its proven strategies and behaviors. The more drastic the change in the organization's environment, the more profound the changes in the organization most likely need to be in order to ensure its survival, its success, or both. Often, the resulting changes are incompatible with the organization's existing perceptions and norms as reflected in its culture and therefore engender conflicts, which must be confronted and resolved. Thus, an organization engaging in double-loop learning questions its basic knowledge and values as well as the existing theory of action. The outcome is new or modified knowledge, norms, policies, and objectives.

Bateson's Type III learning has its counterpart in Argyris and Schön's deutero-learning. In this category, not only is the behavior of the organization subject to change, but the entire process of knowledge generation is questioned as well. Such learning comes from the realization that it is not enough to respond to a particular change in the environment. Deutero-learning requires organizations to learn from previous contexts of learning and to develop organizational knowledge about the learning process itself. In other words, the process of 'learning to learn' is integrated by organizations into their overall values and knowledge base.

Although these categories may act as a guide to studying organizational knowledge acquisition, they have several weaknesses when applied to 'real-world' contexts. The main shortcoming is that they are not always easy to differentiate in practice. They are also highly dependant on the subjective interpretation of a particular event or moment in the life of the organization. One observer may perceive a process to be single-loop, whereas another person may convincingly argue that double-loop learning is involved. In many cases, these forms of knowledge acquisition are not even separate or independent processes; often they overlap in both time and scope.

A further weakness in this tripartite classification is that it focuses too narrowly on problem detection and correction as the primary way of acquiring new knowledge. Although problems, conflicts, and crises are common stimuli of learning, the creation of new knowledge can also be triggered by external opportunities and creative initiatives (Nonaka, Reinmöller, & Toyama, 2001). The categories originally proposed by Bateson (1972) and further defined by Argyris and Schön (1978) are rather limiting in another way as well. They do little to specify how knowledge creation is linked to the external environment - other than suggesting that the environment often serves as a stimulus and that learning is the response.

Despite these weaknesses, it is justifiable to adopt the learning types for analytical purposes. However, it would be desirable to expand and improve the concept by more clearly defining the relationship between learning, knowledge generation and changes in organizational visions and behavior. Table 1 provides a model of this relationship linking a broad spectrum of triggers to the three types of learning. These then result in processes of knowledge generation leading to changes in the organization's vision and thus ultimately causing changes in behavior, structure, culture and potentially also leadership of the organization.

In this process it is hypothesized that visions play a central role in shaping and influencing the different forms of knowledge generation. The addition of overarching visions as a trigger of learning also makes it possible to show that processes of knowledge generation are, in many cases, not confined by organizational boundaries.

Table 1: A model of learning and knowledge generation and organizational change

Triggers		Types of Learning		Process of Knowledge Generation				Outcomes
Crises Opportunities Leaders	→	Single Loop Double Loop Deutero	→	Knowledge Creation Knowledge Activation	→	Knowledge Utilization	→	Changes in: -Vision (strategy, perception, values) -Behavior -Leadership

2.3. Visions

Thus far, I have used the term "vision" without specifically defining it. In the Oxford English Dictionary (Simson & Weiner, 1991) it is described as "a mental concept of a distinct or vivid kind: an object of mental contemplation especially of an attractive or fantastic character; a highly imaginative scheme or anticipation" and as the "ability to conceive what might be attempted or achieved." This definition corresponds with the word's meaning in social theory, where visions have long been understood as a "particular something that orients society's action and behavior" (Papalekas 1959: 224)[1]. Many academics have studied the role that visions play within social systems, particularly their function and relevance within institutions and organizations. Resulting theories have ranged from being rather static to being quite dynamic. In Malinowski's (1944) theory of institutions, according to which institutions adhere to an implicit charter that justifies their existence and survival, the focus was on the stabilizing function of visions. In more dynamic theories institutions are assumed to rest on an idée directrice (Gehlen, 1956; Hauriou, 1925/1965; Willms, 1970) that is said to be mutable, partially indeterminate, symbolically portrayable, yet subjectively unachievable. Schelsky (1970), who specifically addressed visions as an analytical concept does so in the context of developing a theory of institutions that revolved around the phenomenon of social change. He concluded that "the vision of each institution exists primarily in intersubjective communication and [that] every revival or renewal of the . . . linguistic and symbolic interaction in an institution thereby induces . . . change in the institution" (p. 26).

On a more general level, visions serve as frames for collective action, often representing a mental image of some desired future state of the organization (Bennis and Nanus, 1985). Because visions are conveyed as images and metaphors, they are not the same as goals and plans, which are usually more concrete, short-term, and limited in scope. However, this distinction does not imply that visions are too abstract or too broad to be achieved in practical terms. Rogers (1990) noted that "far-fetched conjectures as well as outlandish sci-fi fancies can be distinguished from Leitbilds [visions], mainly in that the latter are considered imaginable and feasible in the foreseeable future" (p. 9).

What visions contribute within social systems is the unique ability to stimulate orientation, motivation, and coordination. Visions are points of orientation in that they are based on core values and shared perceptions. As Mambrey and Tepper (1992) claimed, "a vision can be seen as a collective projection, which is ultimately a synchronization of collective assessments and adjustments" (p. 38). The results of these projections open certain horizons of perception, thinking, and decision-making and preclude others. In this regard, visions orient people and organizations and provide guidance, for they offer a particular perspective that people can either accept or reject as they communicate with each other (Dierkes, Canzler, Marz, & Knie, 1995; Krupp, 1995). The potential of visions to provide orientation becomes especially relevant in periods of change, when familiar criteria for action often lose their validity and people restructure their individual and collective projections of what is desirable and feasible. In this process, visions help

---

[1] All English translations from German language sources are mine

individuals, groups, and organizations relocate their basic goals and values and rebuild a shared perception of the environment.

Visions foster motivation because they activate not only the cognitive potential of individuals, groups, and organizations but also their emotional, volitional, and affective potential. Visions do more than just appeal to the logical and rational mind; they touch upon the internalized norms, values, and preconceived notions underlying people's perceptions, thinking, and decisions. People associate visions with certain sympathies or antipathies, attractions or repulsions, and hopes or anxieties. Thus, visions have the power to motivate people to think and act in a particular direction (Krupp, 1995; Marz, 1993a, 1993b).

Visions facilitate and foster coordination as well, for they are able to mediate between people's different ways of perceiving and thinking (Dierkes and Marz, 1991; Dierkes and Marz, 1994). They serve as vehicles of communication through which to improve the conceptualization and discussion of abstract processes and future options (Dierkes et al., 1996, p. 29). Visions also help people understand complex information and link lay and expert ways of perceiving and thinking. Grin and Grunwald (2000) even went so far as to refer to visions as a "common language" (p. 11) that guides collective action. In this respect visions play a significant role in reducing obstacles to communication and thereby contribute to relatively low-cost and low-loss coordination and cooperation.

Taken individually, the effects of orientation, motivation, and coordination can be quite remarkable. Yet in many cases the three functions are not mutually exclusive. Visions are most persuasive when these functions work and support each other simultaneously. They then have the potential to coordinate the behavior of different individuals, groups, and organizations at different places and sometimes even at different times, becoming powerful tools that often work on multiple levels.

Visions are situated at the mesolevel of singular contexts within an organization, but they also function at the macrolevel - decontextualized societal evolution, participation, and cohesion (Mambery and Tepper, 1992, p. 49). For the purpose of studying the role that visions play in organizational learning and knowledge creation, it is important to distinguish between two particular types of visions: overarching visions in a society at large or in sectors of a society, and visions in individual organizations.

Overarching visions are derived from key elements of the political, social, and economic environment of an entire industry, region, or society. Most of these visions are not restricted to any particular context; they transcend boundaries in order to integrate different contexts. Often, a specific overarching vision is a fusion of several other visions that come together to represent a wider set of normative considerations. Value congruency, ongoing consent, and moral freedom are, as in social contract theory, essential components of these visions. Overarching visions are related to the culture and values of a shared past just as much as they are projections guiding a shared future (Grin and Grunwald, 2000, p. 176).

Visions often vary in form. They describe macrodevelopments, such as the automotive society or the projected advent of artificial intelligence. Overarching visions can also be relatively specific, as is the case with the paperless office or the fully automated factory. They provide the general framework within which individuals and organizations operate.

By contrast, visions specific to an organization encompass the basic rationale for its existence, including its fields of activity, long-term goals, and strategic intents. In addition, they represent the basic values intended to guide the behavior of individual members and of the organization as a whole, the shared perceptions and images of desired membership, and acceptable or unacceptable patterns of behavior.

On the one hand, organizational visions are quite specific. They direct the organization in its efforts to achieve a particular goal and to distinguish itself from others. On the other hand, visions held by organizations are also very general and broad guides that outline an organization's strategy and behavior in changing environments and over long periods. Visions form the core ideology of the organization. Thus, visions held by an organization describe as clearly as possible its future direction and the characteristics that distinguish it from other organizations.

Some visions have been guiding organizations for decades (Rogers, 1990) or even centuries (Collins and Porras, 1994), whereas others, often similarly strong and successful, are ephemeral (Weber, 1996). Such variability signals that visions must be routinely reevaluated and, if necessary, modified in light of changing contexts in order to ensure the future success and survival of the organization. In most cases new or significantly modified visions are stimulated by crises, which indicate a gap between the organization's strengths or core competence and the demands of the environment in which it operates. However, crisis is not always necessary to promote the development of new visions. Strong new leadership within the organization or a profound, credible reorientation of existing leadership can initiate the creation of a new vision or the revision of a familiar one. (see Table 1) One of the main functions of organizational leaders is to explain and, more important, to personify the vision of the organization for other members. Leaders are also responsible for interpreting or reinterpreting the organization's vision in the light of changing environments and for determining the level at which reinterpretation is no longer sufficient. It is in this process that visions guide decisions on what elements of the organization are to be preserved and what kind of future it should be steered toward (Collins and Porras, 1994, p. 66).

Generally, if the values, strategies, and behaviors that are derived from a particular vision prove successful, the organization will pursue them until the need for a new drastic change appears. If such a drastic change is necessary, individuals or groups in the organization will undertake efforts to create a new vision. If the new vision is not successful, another vision will have to be developed from the pool of ideas inside and outside of the organization. This process continues as long as there is the will, the time, and the intellectual and financial resources to be invested.

Organizational visions that have been successful for long periods and that are supported by a fairly stable core membership build a strong culture in the organization. Strong organizational cultures derived from strong visions are a powerful way to minimize the costs of orientation, coordination, and motivation and are a key source of success for organizations (Dierkes, 1988). However, whereas culture describes the actual state of the objectives, values, perceptions, and behavior of the organization as shaped by experience, visions encompass the desirable state.

2.4. Technology and Technological Visions as a Test Case

Technology seems to be among the most powerful of the many forces that are driving change in the environment of organizations. Overarching visions guiding developments in science and technology are therefore important factors shaping the visions of organizations. A wide range of studies on technological trajectories and delphi exercises in nearly all highly industrialized countries strongly indicate that the influence of overarching technological visions and their resulting developments will significantly increase in the forthcoming decades (Fraunhofer–Institut, 1998; Fricke, 1999). The importance of technology for the development and survival of organizations is steadily growing because of the frequency of its use and because of the ever-increasing variety and applications of new technological and scientific developments. Many of these new developments are distinct from their predecessors. They are the result of a profound, but not yet widely discussed, change in the structure of technology, one that is adding a significant new dimension to technological developments and thereby triggering even greater changes in organizations and, indeed, in society as a whole.

The most important change in this context is the shift from artifact-centered technologies to system-centered technologies. Traditionally, technology has been understood as a physical artifact. This view, vociferously articulated by Max Weber (1924), implicitly or explicitly underlies many conceptions of technology in the humanities and social sciences. However, studies on large technical systems (Joerges, 1993; La Porte, 1991; Mayntz, 1993; Mayntz and Hughes, 1988; Weingart, 1989) have increasingly shown that all conceptions in which technology is reduced solely or primarily to physical artifacts and portrayed as a more or less independent sphere of action do not suffice as a framework for explaining the technological changes that these systems represent. To understand, for example, the digital revolution in information and communications technologies, one cannot equate these technologies only with this or that artifact, for their specific nature no longer consists exclusively in physical matter. They are oriented instead to systems applications. The very complexity and speed of this change from artifact-centered to system-centered technologies brings with it a vast range of consequences that organizations cannot escape. An organization that continues to focus on new technologies only as mere artifacts and ignores their overriding nature as systems will find nothing but paradoxes and unsatisfactory explanations as it tries to manage technologies (Bijker and Law, 1992; Jansanoff, Markle, Petersen, and Pinch, 1995; Joerges, 1994; Latour, 1991/1993). By contrast, thinking of new technological developments and visions as systems can pave the way for organizational learning in its various forms.

3. THE RELATIONSHIP BETWEEN VISIONS, TECHNOLOGY, AND KNOWLEDGE GENERATION

Technological visions, new technologies, and organizational visions trigger organizational knowledge generation in three ways:

(1) New or modified overarching visions open up new technological options for organizations and create opportunities for additional ones to develop. In order to profit from these visions, an organization must first be able to perceive their emergence and understand their potential. That ability helps prompt the organization to test the new overarching vision against the organization's existing vision. If the new overarching vision proves to have little or no relevance, the organization may choose not to integrate it into the existing organizational vision. If the vision is considered valuable and the attendant changes minimal, the organization will adapt to the new or modified vision mainly through single-loop learning. This approach leads, for example in business corporations, to modifications in products and modes of production or to other related changes within the framework of the organizational vision.

New or significantly modified overarching visions trigger double-loop learning if they are recognized by an organization but perceived as being so different that they are impossible to profit from within the context of the organization's existing vision. Integration of a new overarching vision into the organization is guided and shaped by the organization's existing vision, which is modified by that very process. This second step of knowledge acquisition also encompasses the creation of new or modified organizational structures and processes, as well as the acquisition of new skills. The more foreign the new technology is to an organization's competencies, the more important double-loop learning becomes and the greater the necessary changes in the organization's vision are.

(2) New or improved technological artifacts or systems become available and affect individual organizations (though perhaps not the overarching vision of an entire industry). Presumably, an organization adopts only those technologies that it perceives to be relevant and superior to its existing operations. If the changes are small, they will be integrated into the organization through single-loop learning, that is, within the framework of the existing organizational vision. However, new technologies that are significantly different from existing ones require double-loop learning that entails changes in the organizational vision. Double-loop learning, consequently brings about changes in the organization's structures, processes, and knowledge base.

(3) The recent shift from artifact-centered to system-centered technologies and technological visions triggers new forms of organizational knowledge generation. The relationship between system-centered technological visions, technological developments, and organizational learning is the most complex and ambiguous form of knowledge generation, for often the visions themselves are still in their formative phase. Moreover, new technologies characteristically entail tremendous, unpredictable change because system-centered technologies are only loosely linked to their artifacts, and these artifacts are accessible to nearly every organization prepared to test new visionary ideas. Organizations therefore have little protection from their competitors and must learn how to respond proactively and quickly to these system-centered technological visions and their technologies.

Additional complications arise because often the system elements of new overarching technological visions may not fit an organization's existing vision. In order to adjust, an organization may first apply single-loop learning in the framework of its existing vision. With more significant adjustments it can also integrate new technological visions into its own vision through double-loop learning. In most cases, however, these two options will not be sufficient to deal adequately with the complex changes introduced by system-centered technologies and their visions. Organizations, instead, have to question their single- and double-loop learning patterns and search for a different mode of knowledge generation. In this environment, learning to learn, or deutero-learning, is a necessary element of the organization's vision and culture. This third form of learning leading to knowledge generation, which is always mentioned in classifications but seldom described and analyzed in the reality of organizational life, is often central to survival and success in system-centered fields of technology.

4. TECHNOLOGICAL VISIONS AS TRIGGERS OF AND OBSTACLES TO ORGANIZATIONAL KNOWLEDGE GENERATION: LESSONS TO BE LEARNED FROM CASE STUDIES

To illustrate the complex interaction between overarching visions and technological developments, organizational visions, and organizational knowledge generation, three case studies have been selected that have a common background in the change in information and communication technologies since the 1960s. (For an overview see Table 2) The focus is on the vision underlying this change, the computerization of the everyday world. In all three cases, scientific breakthroughs led to new or significantly modified technological visions that were either perceived and used by organizations as an opportunity to learn or that were not seen as important enough to stimulate the generation of new knowledge and thus motivate a change in the original vision of the organization.

Table 2: Case Studies

Case Studies		Overarching Vision		Organizational Vision		Dominant Types of Learning		Results
Word-Processing Company	—	Changing	—	Old	—	Single-Loop Learning	→	Stagnation
Automotive Industry	—	Modified	—	Modified	—	Double-Loop Learning	→	Stagnovation
Internet	—	New	—	New	—	Deutro-Learning	→	Innovation

The first example is drawn from the word-processing industry. This case shows how an organization, instead of adapting to a new overarching vision through processes of knowledge generation, chooses to maintain the status quo, halting the creation of additional knowledge and ultimately causing the organization to become outdated in the face of changing technologies and markets. The second example, from the automotive industry, presents a combination of both stabilizing and innovative reactions to new environmental challenges. Changes within this industry have shaped the visions and behaviors of organizations in a way that has led them to press forward with multiple forms of knowledge generation, simultaneously questioning and challenging traditional technologies, yet ultimately preserving them. The final example, the Internet industry, shows the diverse outcomes that can result from two quite different approaches to the same technological challenge. In this industry technological innovation has been, and still is, so fast and revolutionary that organizations pursuing single- and double-loop strategies quickly become isolated and then left behind. However, the organizations that proactively undergo deutero-learning become the leaders and shapers of new technology within the industry.

4.1. The Vision of the Word-processing Industry: Stagnation of Learning

The first example, the vision of word-processing technologies, is typical of what happens when an organization adheres overwhelmingly to a particular vision. This case is that of AEG Olympia Büromaschinenwerke, a German office-equipment manufacturing company founded in 1903 and dissolved in 1994. In this organization "the mechanical typewriter [was] the vision in which generations of employees found meaning" (Buhr, 1997, p. 55). Both management and employees at all levels and in all departments were imbued with this vision and had "unswerving loyalty to the typewriter as a product" (p. 56). From its inception, the typewriter was the core of Olympia's technological vision, and its subsequent success defined and shaped the company's corporate culture. Managers were heard to say "I wouldn't like to have to see the day when we no longer make typewriters here!" (p. 57). Accordingly, all the activities of thousands of employees revolved around this one product and the aim of perfecting it. It encompassed the projections of what the entire organization held as desirable and feasible, and the employees thrived on this vision.

However, in the mid-1960s signs indicated that the typewriter, the classical form of mechanical writing and the underlying central vision, would eventually be supplanted by a new technological vision, electronic word-processing systems (Bammé et al., 1983; Hofmann, 1997). The company's strong identification with the original vision began to limit the organization's ability to perceive these changes in the environment and ultimately proved to be a debilitating obstacle. The fate of two internal attempts at innovation graphically illustrates how the success and power of Olympia's vision constrained further knowledge acquisition and trapped the organization into maintaining an increasingly obsolete status quo regarding its core technological vision and knowledge.

In 1971 the board of directors installed a project team called "nonmechanical writing." It consisted of external engineers, "totally new people who would be able to see the company's problems and who had no preconceptions" (Buhr, 1997, p. 58). After first experimenting with different alternatives derived from the new overarching technological vision, this group concentrated on developing a specific component of the new electronic word-processing system, the ink-jet printer. Although the team had very limited resources and although its work in the company was at best only tolerated and was usually met with overt or covert resistance, it succeeded in developing a functional prototype.

Real conflict flared, however, when this new product confronted the traditional, typewriter-centered organizational vision of the company. At the meeting in which the ink-jet printer was presented, it failed to gain the necessary acceptance primarily because it departed too greatly from the organizational vision and the mechanical writing technologies to which the audience was so accustomed. The engineers of the development team tried to reconcile Olympia's employees and management to the new machine by attempting to "somehow make a typewriter" out of the printer (Buhr, 1997, p. 60). For example, because with typewriters each letter is immediately visible after the corresponding key is struck, they rebuilt the ink-jet printer in such a way that its continuous operation was interrupted after every strike of a key so that the letter could immediately be seen. To accommodate the demand for carbon copies, the engineers used special paper that could be permeated by the ink. However, neither modification was able to change the deeply rooted attachment that management and the employees had to the typewriter-centered vision, and as in most traditionally structured organizations, the development engineers lacked the necessary power to impose such a change.

A second attempt at innovation, which was launched in the late 1970s, was similarly futile and only confirmed that the company's vision of the typewriter was indeed entrenched. In this effort another small team of engineers went about developing a computer-aided word-processing machine. To promote the acceptance necessary for their project, they adopted the typewriter-centered vision from the outset and tried to create "a hybrid between a typewriter and a personal computer (PC)" (Buhr, 1997, p. 61). Unlike the ink-jet printer, this innovation was presented to Olympia's employees and management as a typewriter from the beginning, not retrospectively. The development group built the new hardware and software around the old, but widely loved, product so that "the typewriters equipped with a display [were] a kind of PC, except that you [didn't] see it" (p. 61). These machines constituted an attempt to anticipate and tie into the new overarching technological vision of computer-aided office communications. Yet it too failed. The PC-disguised-as-a-typewriter was accepted by neither the company nor its customers.

There were no further efforts to adapt Olympia's vision to the new overarching vision in word-processing technology. Despite the technological changes taking place throughout the industry and in the world, the company unwaveringly continued to develop and improve classical typewriters in the firm conviction that they would sell far into the twenty-first century (Buhr, 1997, p. 59). However, as technology progressed, the former leader in word-processing technologies became an outsider to technological change in the industry - and then its victim.

4.2. The Vision of the Automotive Industry Challenged: Stagnovation as a result

The vision of the automotive industry is a second example of the tremendous power and depth that can be associated with technological visions. However, whereas the vision of the typewriter was rooted mainly in a small number of particular organizations, the vision of mobility - centered around the automobile as the most desirable form of transport - is rooted in many organizations: automotive firms and their suppliers, police organizations and courts, governments, unions, and areas outside the organizational sphere, such as the daily behavior of drivers and their individual and collective projections of what is desirable and feasible (Canzler, 1996; Canzler, Knie and Bertold, 1993; Canzler and Marz, 1996). This vision, which has been the foundation of the automobile's dominance in all notions of mobility, has structured transport policies throughout the world for decades and must be one of the most widely held, stable, and effective overarching technological visions ever known (Dierkes, Canzler, Marz and Knie, 1995).

The vision of mobility centered on the automobile is currently so powerful that it pervades the learning of almost all organizations in the industry. Thus, the growing crisis of mobility, as indicated by traffic jams, pollution, and the threat to global resources, is treated as though it were merely a variant of other problems that have already been successfully managed. This kind of learning is fixated on retaining the time-tested, technological core of the automobile - the "high-performance, internal combustion limousine" (Canzler and Knie, 1994, p. 61) - but on upgrading its equipment and accessories to meet society's changing needs. Sensory, optoelectronic, and information technologies are being used by the automotive industry in the great hope of prolonging the original vision of the automobile. Building upon the traditional technologies of the automotive manufacturing and upon technologies hitherto having nothing to do with cars, these innovations show that knowledge acquisition has gone beyond the original vision of the industry. The networked and intelligent car is the central technology now being propagated in a revised and reshaped vision of the automotive society of the future.

Although the degree to which additional technological input can enhance the overall efficiency of the automobile-centered transportation system has been occasionally overestimated, there is no reason to assume that its potential capacity for solving problems is minor. Specifically, a vehicle fully equipped with modern sensory, optoelectronic, and information technologies is expected to bring three kinds of relief from the negative impacts of the automotive society (Canzler, 1996). First, collective information systems reporting on changes in traffic conditions and parking availability and satellite-aided global-positioning systems have been introduced to improve the efficiency with which the existing infrastructure is used. Second, the continued development and integration of information technology is intended to lead to pretravel information systems. Available on-line through personal computers, they are to aid the planning of journeys by expanding the quality and amount of information available to drivers before they depart. An unprecedented level of efficiency in one's planning seems imminent. Third, technological developments have also made it possible to control the volume of traffic by electronic road or congestion pricing. Electronic tolls and restrictions are regarded as a way of introducing market principles into the use of transport infrastructure and of thereby helping minimize undesirable social and environmental side effects.

This process of sustaining the power of an overarching technological vision by modifying it through technological components from other fields of technology can perhaps best be described as stagnating innovation, innovative stagnation, or for short, stagnovation (Canzler and Marz, 1996, 1997). This expression communicates that the decade - old vision of the automotive society has been neither preserved nor completely reshaped or replaced. Bringing new information technologies into the high - performance limousine creates both virtual and real space for innovations. However, these innovations are ultimately seen as stagnating because they do not transcend the traditional vision of the automobile as a way of satisfying the human demand for mobility, which is likely to grow significantly. Relieving the problems of the automotive system only temporarily will most likely postpone the collapse of the transport infrastructure in metropolitan areas, but it does nothing to eradicate its underlying problems.

4.3. The Vision of the Internet Created and Recreated: Learning to Learn

The classical vision of the Internet rests upon a commitment to an open architecture of transmission, a design that distinguishes this network from other communications systems such as the telephone (Dierkes, Hofmann, and Marz, 1998). Most of the scientists who originally developed and worked on the Internet's transmission principles were based at universities. It was their projections of what was desirable and feasible that gave the Internet vision its start (Comer, 1991; Huitema, 1995). The culture of the academic community, where research results can easily be published and discussed and where knowledge is shared and made readily available to others, was mainly responsible for shaping the initial vision and culture of the Internet (Dierkes et al., 1998).

The Internet's complex structure and dependence on system-centered technology is in many ways unparalleled, particularly in its vision (Dierkes et al., 1998). The founding principle of the vision as stated in Requests for

Comment (RFC) (Bradner, 1996) - that development of Internet standards shall be open - means that anyone with the necessary expertise is invited to contribute (Helmers, Hoffmann and Hofmann, 1997). Design standards, known as Internet drafts, are published, discussed, and then archived in specific series, such as RFC, which are available to all users. Thus, the work on developing Internet standards is a collective effort regulated not through affiliation with a particular organization or committee but instead solely through "the excellence of the technical work" (Huitema, 1995, p. 24). If a pool of internet drafts contains two competing designs, and if it is not clear which design is superior, each is developed by a work group until ready for implementation. The activities of the different groups are transparent and open to objections, critique, and suggestions. The only prerequisite, or running code, for acceptance of a new standard is that it have at least two independent implementations proving that the technical specification in question functions and that it is interoperable with the other technical constituents of the Internet (Alvestand, 1997; Bradner, 1996). The extent to which a standard then prevails is not subject to the control of specific organizations. The market decides instead. The environment provided by this vision and by the resulting culture of the Internet is highly flexible, cooperative, and decentralized but difficult to manage and regulate.

Not only is the process of developing standards open, but its results are open as well. The Internet's specific transmission principles, called protocols, are not standards in the formal or conventional sense. These protocols are not formally decided upon, introduced, or declared binding in any manner. Instead, they are passed on and offered for discussion, both voluntarily and without cost. In fact, the Internet is as much a collection of communities as it is a collection of technologies, and its success is largely attributable to its record of both satisfying basic public needs and using the community as an effective tool to push the Internet forward (Cerf, 1998). Consequently, the speed and transparency of Internet development has been unparalleled.

However, in the early 1990s three parallel processes confronted the Internet's founding groups with a completely new environment. First, government increasingly withdrew its funding as investment from the business community dramatically increased. The centers for the development of standards thereupon shifted away from universities and the academic community to the major users and providers of computer and network technology. As a result, more and more researchers who were committed to the open development of standards migrated from universities to the business sector (Lehr, 1995).

Second, the World Wide Web and its browsers provided new Internet uses whose user-friendliness attracted a mass public able to operate within the system without any training or expertise. This explosion in the number of users not only dramatically increased network traffic (Helmers, Hoffmann, and Hofmann, 1996) but also radically changed the methods and content of daily use. Recent figures show that the number of worldwide Internet users doubles every hundred days (Weingart, 1989). Clearly, the classical academic users have lost their preeminence.

Third, this quantitative growth and qualitative shift in the community of users led to the discovery of the Internet as a viable market. Consequently, numerous Internet-specific firms were quickly joined on the net by more and more traditional companies and organizations seeking market exposure and an opportunity to offer their services on-line (Dierkes et al., 1998).

The general shift from having a few networks with a modest number of hosts to having many networks has resulted in an entirely new Internet environment, one in which the development of standards has become more crucial than ever. Although the overarching vision remains the same, some groups within the Internet community have begun to challenge the net's traditional methods of standardization. In general, two different approaches to standardization have surfaced; those represented by the Internet Standards Organization (ISO) and those of the Internet Engineering Task Force (IETF). These organizations have some practices in common: Both of them advance technology through a committee's effort and a process of consensus, and both use some form of parliamentary procedure.

In many other respects, however, particularly in their history, culture, and vision, these communities differ substantially (Piscitello and Chapin, 1993, p. 13). The ISO, for example, tackled the new and fast-paced environment of the Internet industry by applying the traditional vision, strategies, and technologies of their organization's specific competencies. The ISO is a worldwide federation of national standards bodies from 130 countries (http://www.iso.ch). Its scope is not limited to any particular industry; it is well established in such diverse fields as information-processing and communications, textiles, packaging, distribution of goods, energy production, and banking and financial services. At the time of Internet development, the ISO and its study group, the International Telegraph and Telephone Consultative Committee (CCITT), had been working closely on the development of public packet-switching standards to coordinate efforts to develop a single international reference model for open-system interconnection (Piscitello and Chapin, 1993, p. 14). Having this experience in standard-setting, the ISO saw the Internet as a related field of technology and tried to use mass support by its member states to set new standards and control basic technological elements in order to gain influence over the Internet (Dierkes et al., 1998). Although many dimensions of the basic technology of the Internet were familiar to and shared by the ISO, the body was a closed and traditionally structured standard-setting institution, so its culture and standard-setting procedures were dramatically different from those common to the Internet. The ISO failed to recognize the open, system-centered nature of the Internet and thus did not adapt its vision and procedures to the specific Internet culture. Ultimately, by trying to transfer to the Internet the artifact-based technologies and practices of standardization that had served well for decades in other related fields, the ISO defied the vision of open standardization practiced up to that time (Dierkes et al., 1998). As a result, most of the standards developed by the ISO never reached the point where they could be applied to commercial products in the field.

The IETF took a different course. As an independent institution without legal status or formal authority, the IETF had a culture rooted in the original vision of the Internet - the open development of standards - and was championed by work groups composed of engineers and scientists from academic, computer, and telecommunications communities. These work groups were more flexible and fluid than other standard-setting bodies. They were often composed of a diverse, ad-hoc group of volunteers. They approached the development of Internet standards as a new experience and an opportunity to learn and proactively manage the changing nature of their environment, not just as a mere variant of already familiar tasks. This culture and vision enabled the IETF to become the avant-garde in developing standards openly (Lehr, 1995).

Although various companies now employ many of the same individuals who had been actively involved in the development of the IETF, the vision and practices modeled by this organization have endured. Many companies of the 1990s embrace the overarching vision of the open development of Internet standards and grant their former IETF founders the necessary scope for work in this field (Dierkes et al., 1998). These organizations allowed a major share of these employees' available working hours to be dedicated to writing Internet drafts, reading corresponding mailing lists, and discussing current problems. In addition, these companies sent large delegations to IETF meetings, attendance at which had come to range between 1,500 and 2,000 participants. Despite their power and influence in the Internet industry, these companies did not try to turn active IETF members into corporate representatives. Any attempt to do so would have been counterproductive because high-technical competence and affiliation with the IETF elite was defined in part by the ability to resist such influence peddling. Esteem and recognition went to those who strictly distinguished the vision of their specific organization from the overarching vision of the Internet as a collective good.

5. WHAT DO THESE CASES TEACH US ABOUT VISIONS AND KNOWLEDGE GENERATION?

The three case studies from the word-processing, automotive, and Internet industries show that visions can be either a trigger or a barrier to the acquisition of new knowledge (Berthoin Antal et al., 2001). They also demonstrate why it is necessary to increase attention to the task of developing organizational visions as a mechanism for knowledge generation.

The case of Olympia demonstrates how organizational knowledge generation can become so constrained by an organization's particular vision and perception of technology that even unmistakable changes in the overarching vision cannot persuade a company to shift to a different type of learning. The knowledge creation in this organization was basically confined to single-loop strategies within the framework of the existing vision. The problem was not that the company lacked the capacity to innovate or meet the technological challenges on the market but rather that the typewriter program and its decades of success proved so powerful and deeply embedded in the organization that this legacy blocked management's view to new technologies on the horizons.

Olympia's ability to generate the necessary knowledge base was constrained by boundary rootedness. This boundary reveals the extent to which overarching visions are accepted and anchored in the organization's vision. If a vision is embraced across an organization and even beyond, it takes a long time and unique circumstances for the organization to recognize changes in the overarching vision. Instead, the organization will have a tendency to confine itself to single-loop learning. Coupled with existing leadership and a strong culture, this response may improve existing products and strategies, but the organization will find it nearly impossible to modify its vision and to develop other strategies of knowledge acquisition.

This conclusion is supported by a study that Collins and Porras (1994) conducted on the successful habits of visionary companies. They argued that organizational visions must transcend existing products and practices or they can easily become obsolete. If visions are shortsighted or centered on a specific product, they can actually prevent organizational knowledge from going beyond incremental improvements based on single-loop learning: "[B]y confusing core ideology with noncore practices, companies can cling too long to noncore items - things that should be changed in order for the company to adapt and move forward" (p. 81). Over time, the goals, strategies, products, structure, and culture of an organization will most likely need to change in response to the environment. Thus, visions must be flexible, or organizations must be willing to change their vision in order to accommodate the various stages of its life cycle. Although the core ideology of visions must be specific, organizations must separate these core values and goals from their present products and successes. Collins and Porras described visionary companies as those that are able to instill their core ideology into their vision but that are simultaneously prepared to reevaluate and, if necessary, modify other aspects of the existing vision.

The example of the automotive industry portrays many of the same characteristics as that of the word-processing industry: mainly, an industry's or society's dominant, overarching vision and strong visions of new technologies emerging in individual organizations and related fields. However, the potential of these visions and technologies triggered the acquisition of new organizational knowledge throughout the industry.

Moreover, knowledge generation has extended beyond the original organizational vision and single-loop strategies; double-loop learning was apparent. Organizations modified their specific visions and learned how to improve and modernize by exploiting the given, albeit narrow, scope of action within the overarching vision. Borrowing technologies from other industries by bringing computerization into the vision of the high-performance limousine has opened both virtual and real space for double-loop learning and hence has also modified organizational visions.

The example of the automotive industry illustrates the controlling power of a firmly established overarching technological vision. It has been adopted not only by an entire industry but by society as well to such a degree that the vision has become virtually frozen in time, or at least stagnant. Such stagnation is often the result of barriers to the acquisition of new knowledge by organizations. Organizations operating within this framework are confronting barriers of control and substance. The barrier of control results from the attempt to take the process of shaping an organization's vision and steer it in a particular direction. In principle, very little, if any, such control is possible. There is one factor, however, that can be powerful enough to influence any vision directly: organizationally perceived change in the overarching vision of the industry. Yet, in order for an organization to respond, it must first perceive this change and then recognize a need to adapt its vision accordingly. Despite current arguments claiming that the automobile is currently pushing the limits of its social desirability, the power and control exerted by the vision that enshrines it has blocked drastic changes or alternative visions from developing.

This inability to look beyond the vision and to the horizon of emerging technologies can stifle the development of new visions and technologies within organizations and thereby hinder the generation of new knowledge. Although many organizations within the automotive industry have recognized the need for profound change, the development of new visionary substance has been limited because learning has been confined to single- and double-loop strategies. Organizations within the automotive industry have been able to surmount or at least minimize this barrier of substance by concentrating on modifying their existing vision. The visions, strategies, and culture of this industry continue to focus on the traditional overarching vision of the automobile. But in order to ease the growing problems associated with the automobile, individual organizations have successfully incorporated new technologies from other fields into their own specific visions. Some of this knowledge acquisition has transcended organizational boundaries, and new visions within organizations have triggered incremental changes throughout the entire industry. Drawing on Collins and Porras (1994), one could say that the automotive industry has been able to preserve its core yet still make progress. However, as the challenges of the automotive society intensify, the human need for mobility will require both the elimination of these barriers and a high level of preparedness to generate new knowledge within organizations and the industry as a whole.

The final example, the Internet industry, demonstrates the impact that organizational visions have on single-loop and deutero-learning. The ISO confined its acquisition of new knowledge to single-loop strategies because it failed to expand its understanding of technologies beyond the traditional, artifact-centered notion of technological development. The ISO's tendency to be political led the organization to embrace multiple solutions to a single problem and to show little concern for developing a real understanding of the underlying technology and its general application (Piscitello and Chapin, 1993, p. 26). The ISO tried to apply to the Internet the same strategies that had brought it previous success in other contexts and different environments. Although a focus on the artifact dimension might have been sufficient to facilitate the construction of global telephone services and standards for other artifact-centered technologies, the rapid changes and developments in system-centered technologies easily overloaded this organization. The ISO's failure to align its vision with the overarching vision in the industry ultimately blocked new visions from triggering the degree of knowledge generation necessary to operate within this industry.

The IETF, by contrast, perceived the systems nature of this technology and ensured that its strategic framework coincided with the technological vision and culture of the industry. Parting with the traditional approaches to technological change, the IETF has pressed forward with Internet developments through an innovative kind of decentralized, cooperative, and open coordination rather than adhering to a centralist, hierarchical, and closed organizational model. This more flexible approach promoted permanent learning as an important element of the organizational vision and has thus allowed considerable latitude for adapting the development of standards to the new system-oriented technology that emerged (Piscitello and Chapin, 1993, p. 27).

There are many indications that this form of innovation, the open model for developing technology and standards, emerged from a kind of deutero-learning emancipated from servitude to past programs and experiences. The IETF illustrates how an organization can overcome what can be called the boundary of acceptance. The boundary of acceptance stems from the degree to which the organization's members are involved in the process of creating a new organizational vision. The more they are engaged in the process, the lower the barriers are; the less the members are engaged, the higher the barriers. When a vision is developed without input from organizational members, its power to orient, motivate, and coordinate is usually never brought to bear, so its effect on the organization's ability to acquire new knowledge is rather limited.

In the IETF, vision development was unique in that the stakeholders of the technology - the researchers, users, and vendors - were those directly involved in creating and shaping the vision. Membership in the IETF was intentionally left open so that users would be reckoned as part of the actor constellations that help propagate socially desirable innovations (Dierkes et al., 1998). Open involvement in shaping an organization's vision can facilitate its acceptance and allow members to collectively acquire new knowledge and make the transition from single- and double-loop learning to deutero-learning. The example of the IETF offers one piece of evidence that developing a successful vision does not require any special form of guidance or control, or even charismatic leadership. If constantly striving for new knowledge is deeply rooted in the organizational vision and, over time, in its culture, the vision will then belong to the entire organization, empowering and stimulating the learning of all its members.

6. RETHINKING THE CONCEPTUAL FRAMEWORK: THE FIRST CONCLUSION

Review and comparison of these examples of different forms of knowledge acquisition reveals three main relations between visions and the generation of new organizational knowledge using changes in technological visions and technologies as triggers. The first relation has to do with the dimensions of technological change and the way they are perceived. Not all organizations have the organs of perception needed to recognize changes in an overarching vision and sense their potential. Even those organizations that do have such organs or sensors often lack the collective "intellectual" capacity needed to integrate perceived changes into their organizational vision. Furthermore, organizations cannot respond to each and every change and development in their accustomed fields of technology, let alone those in related ones. The opportunities introduced by new technologies are overwhelming, and organizational resources allocated to innovation are often limited. Organizations are therefore forced to decide which options to explore.

In general, organizations welcome new or modified technological visions and new technologies if they can be easily managed within the existing framework of their organizational vision. Even drastic technological changes at the artifact level can still be perceived and responded to with relative ease by organizations, for most of these changes do not disrupt the culture or the operational predictability of the organization's activities. However, the rapid growth of system-centered technological change has challenged this complacency. In the long run, it can be highly problematic for processes of knowledge generation if organizations recognize or concentrate on only the physical dimension of technological change to the exclusion of the other dimensions. System-centered technological changes require organizations to reconceptualize both their goals and their work activities, or, as Bloomfield and Vurdubakis (1997) referred to it, the "life world of their professional practices" (p. 647). If an organization does not respond in such a manner, it will often be unable to meet the challenges posed by rapid system-centered technological change, a failure that can ultimately prove fatal.

The second relation between organizational knowledge generation and visions is the influence that the different types of learning have on how organizations deal with change in technological visions or developments entailed by new technologies. Organizations are often unprepared for, or do not have the capacity to deal with, the many issues raised by changes in technology. The reason is that technological change can erode members' confidence in what they know and the way in which they have acquired their knowledge. This dynamic is especially characteristic of system-centered technologies, which until recently were still regarded as the essence of science fiction, not everyday reality (Shillingford, 2000). Whether and how much an organization is able to adapt to these complex technologies depends greatly upon the type of knowledge generation to which the organization can resort. Although single- and double-loop learning can sometimes be sufficient even for significant improvements, time-tested programs and experience will do little, if anything, to help organizations adequately perceive and proactively deal with system-centered technological changes. This deficiency of single- and double-loop learning is especially apparent in industries where major contextual changes are so rapid and revolutionary that they can only be described as discontinuities.

In general, organizational visions and cultures in which learning is valued as an essential and permanent element of practice and procedure are where corporate strategic flexibility and the use of new technology will increase (Hitt, Keats and DeMarie, 1998, p. 35). Organizations with such visions and cultures are capable of creating processes of knowledge generation beyond single- or double-loop learning. They recognize that technological change has forced the context of organizational learning itself to undergo profound changes, a process that, in turn, calls hitherto successful learning strategies into question and requires development of the capacity to learn and relearn. Although deutero-learning does not guarantee an organization's success, it may well improve the chances that the organization will be able to adapt to fundamental changes in technology.

The third relation brings the various contexts of learning, and visions in the field of technology together to show the power that visions have in shaping organizational knowledge generation in general and each of the learning types in particular. As all three examples illustrate, this power can be extraordinary. In some organizations, visions can be so strongly bound to existing technology that the acquisition of new knowledge is precluded because profound changes of context are either unperceived or are defined as irrelevant and ignored. These types of organizations pursue single-loop learning within the boundaries of existing organizational visions and often stagnate because they are unable to innovate and adapt successfully to new overarching visions. Donald Sull (1999) referred to this condition as "active inertia" (p. 42), the organization's tendency to follow established patterns of behavior even in response to dramatic environmental shifts. In such cases there is a tendency to force new visions and technologies into existing schemata. Although long-standing visions and success often provide stability by giving organizations a strong, unified sense of perceptions and values, they also make the organization and its members reluctant to change. Despite incremental innovation in products and services, these organizations continue to work with old, even outdated processes and structures. Instead of reevaluating strategic frames in light of new overarching visions, they simply accelerate their tried-and-true activities without any real concern for the changing nature of their environment.

The ability to perceive and integrate new knowledge requires organizations to scan both their internal and external environments and develop visionary substance and long-range strategies to encompass new goals, values, and behavior patterns. The higher the value that learning has in an organization's existing vision and culture, the more likely it is that the organization will be able to perceive and exploit opportunities for innovation offered by new overarching technological visions, even those quite remote from its current field of activities.

Each type of knowledge generation is based on a different relationship to organizational visions. Visions thoroughly shape forms of learning within an organization. Yet visions, too, are the result of a learning process. They grow from a social field of similar, preexisting, intertwined projections of what is desirable and feasible. Visions are a point of orientation enabling people and organizations to specify and fine tune their individual and collective projections of what is desirable and feasible, a process that stabilizes and reinforces these projections. Visions, therefore, are both a result and a trigger of learning. For this reason a change in an organization's process of knowledge acquisition, especially the transition from single- and double-loop learning to deutero-learning, is extremely difficult, if not impossible, without a change in the original organizational vision in which these types of learning are anchored.

Thus, visions are a powerful instrument for guiding organizations. By critically reflecting upon both organizational and overarching visions, organizations have a tool with which to anticipate and prepare for future developments. Those visions that rest on the core values of the organization and allow for a degree of openness and flexibility can enhance the responsiveness of organizations and can help them ultimately overcome the problems of active inertia. However, the power of visions to trigger the generation of new knowledge depends upon the degree to which learning is integrated into the organizational vision. Organizations that are proactive and able to develop a pool of new potentially visionary ideas and mold them into a viable and widely acceptable vision will have a major advantage in the current state of rapid and unpredictable change.

7. A RESEARCH AGENDA ON KNOWLEDGE, LEARNING, AND THE ORGANIZATION: THE SECOND CONCLUSION

Visions are only one, though a very important, element in the process of an organization's effort to generate new knowledge. The ability and willingness of individuals to learn and the culture, structure, and leadership of organizations are among the other factors affecting creation of new organizational knowledge by either promoting or blocking processes of screening information and of deciding whether they should be transformed into organizational knowledge (Berthoin Antal, Dierkes and Hähner, 1997). Technology, though very powerful, represents only one of the many triggers stimulating the acquisition of new knowledge in an organization. Markets, competitors, business partners, governments, and social movements are other such factors (see section II in Dierkes et al., 2001).

However, factors that shape processes of knowledge generation within organizations interact in many ways. Their complexity is revealed by in-depth analysis of visions as enabling factors for knowledge creation and by analysis of technology as a key trigger. Such study also illustrates that current academic understanding of these processes is still rudimentary. What is relevant in research on this subject? What aspects should receive priority? The answers to these questions are quite subjective. However, the illustrative treatment of visions and technology can be used to help deduce clusters of research questions; to clarify issues; to pose worthwhile questions for study; to hone current expertise in knowledge management by transferring knowledge from other fields, particularly from the study of organizational learning; and to introduce unknown territories requiring completely new research initiatives.

7.1. Conceptual Aspects

In the overall field of research into the knowledge-based organization and knowledge management, a crucial task is to come up with an operational distinction between information and knowledge (Berthoin Antal et al., 2001). A great deal of literature, especially publications about new information and communication technologies, does not distinguish clearly between information, which these technologies make available to a growing range of individuals in the organization, and knowledge, which simply can be seen as information relevant in a specific context. Many devices discussed in the literature deal with information and not with knowledge. Therefore, a key conceptual challenge is to design processes for translating information into knowledge.

Another is to develop a taxonomy of knowledge, as a basis for ascertaining the links between the existing knowledge base, the process of organizational learning, and the new or extended knowledge base of the organization. Basic studies of the knowledge base as organizational memory are also important for gaining an insight into to the extent to which knowledge is stored in individuals, groups, files, and culture and to how these different modes of storage are effectively tapped for problem-solving and decision-making. The more important the contextual dimension of organizational knowledge is, the more important such a distinction between information and knowledge becomes.

A key element of such a taxonomy is the question of the difference between tacit knowledge and explicit knowledge and the closely linked question of how to transfer tacit knowledge to explicit knowledge, if it is deemed necessary, or how to transfer tacit knowledge from individuals, groups, or whole organizations to other individuals, groups, or whole organizations. This question is hardly new. Many fields such as surgery, leadership, the crafts, and the arts are deeply rooted in tacit knowledge. For centuries, a prime concern of educators and leaders in these sectors has been to find effective ways of transferring this knowledge from one individual or group to another. Because the knowledge base of individuals is larger than what is normally used in the organization, the question is how much of an extended knowledge base is really relevant for specific contexts of problem-solving and decision-making or strategic choices in an organization (Nonaka et al., 2001a)?

7.2. Factors Influencing Knowledge Use and Creation

7.2.1. Structure and Knowledge

In this part of the proposed research agenda, four questions are at the forefront of inquiry:

(1) What kind of structural devices enhance the use of existing knowledge in the organization?
(2) How effective are they?
(2) What kind of devices are able to extend the existing knowledge base by drawing on external sources?
(3) What are the main barriers to using the knowledge base, and how can they be overcome?

This is where the above-mentioned 'chief knowledge officer' comes in as a structural device. Other relevant structural elements are the 'company university'; flat organizations; a degree of convergence between various layers of management, task forces, and temporary teams, and the role of boundary-spanning personnel, consultants, and boards. However, key decision-makers may perceive an unmanageable volume of information constantly flowing into the organization and floating around in it. This understanding raises two additional questions: What are the key structural devices for selecting the relevant information from the irrelevant and the important from the less important in order to avoid information overload? What are the structural obstacles to the use of existing knowledge? These are important research issues linking structure to culture, the overall organizational emotional climate, and the question of power and its use. An organization that behaves as though "power is the right not to learn" is significantly limiting its ability to tap into existing knowledge. Organizations with long hierarchies may lose a great deal of knowledge or may distort knowledge as it moves from its source to key decision-makers. Compartmentalization of an organization may be another important impediment to the creation of new knowledge.

On the other hand, relevant knowledge can be lost through information overload just as easily in a very fluid organization with many teams and task forces but little coordination. Under such circumstances, key decision-makers are confronted with too many readings about the environment, too many strategies for reacting to environmental changes, and too many suggestions about how to pursue which alternative(s).

Structurally, links between various parts of communities of practice or knowledge cultures also have an important bearing on the creation and use of knowledge. Some companies try to manage R&D by linking it very closely to marketing, production, or purchasing, whereas companies are convinced that an R&D function that operates in relative isolation from day-to-day pressure ensures more creativity and can therefore increase the innovativeness of solutions (Berthoin Antal et al., 1999).

7.2.2. Culture and Knowledge

Culture is one of the central storage devices of organizational knowledge. Standard operating procedures are one example (Kieser, Beck, 2001). So are shared success-related perceptions shaping the selection of different knowledge bases of functional areas, other subcultures, or professional communities. A key question in terms of knowledge management is how to create an organizational culture that fosters both broad use of the existing knowledge base and open access to outside knowledge. Research suggests that the process of condensing information into knowledge is greatly influenced by culture, defined as the internalization of past successes (Berthoin Antal, Dierkes and Hähner, 1997). It seems that organizations tend to be inherently conservative in the process of converting information into knowledge. The question therefore is how to avoid this "dark side" of a strong and successful culture and instead allow different information and alternative ideas to enter into an organization's knowledge base, where it can be used as a resource for strategic choices, problem solving, and decision-making. Obviously, in rapidly changing environments and in an era of high unpredictability, a culture has to be very open to learning, unlearning, and the inclusion of new, unfamiliar, or underutilized knowledge bases. The central question therefore is how to ensure the development of such a culture? Can it be managed? Or does the development of a culture just happen to a company in the course of its history as it experiences the importance of learning and of storing that learning deep in its culture as a primary source of its knowledge?

7.2.3. Leadership, Visions, and Knowledge

Leadership plays an important role in interpreting the relevance of knowledge for the success of a company (Sadler, 2001). There is a close link to be hypothesized between leadership and the role of visions in knowledge-based firms. A key task of leadership is to create a vision, interpret an existing vision, and then initiate processes of revising the vision of the organization in times of rapidly changing environments. In such periods, it is important that constant learning and an openness to new knowledge bases be integrated into the culture of the organization. During rapid and unpredictable change, it seems to be necessary to create a common understanding of the importance of permanently questioning and unlearning obsolete knowledge while adding to the existing knowledge base of the organization. Leadership has to live up to this vision of a knowledge-creating organization. Reward systems must support members who follow up on such a vision.

7.3. Technology and Knowledge

A significant part of the literature on knowledge management and the knowledge-based organization stresses the importance of technology, particularly network technology that allows rapid distribution of information to the largest possible number of people. Several questions are of interest in this research: Which media are really relevant to the transfer of information and knowledge (Büchel and Raub, 2001)? How much physical presence can be replaced by technology in the exchange of knowledge and processes of accumulation and aggregation of knowledge? Does more accessible information provided by data floating around in the organization really enhance the knowledge base of the organization or does it just keep its members busy processing information without their seeing its contextual relevance for problem-solving, decision-making, and strategic choices? This part of the literature is often highly abstract and occasionally oversells the role of information and underestimates the contextual dimension of knowledge.

Thus far, it also seems to be fairly weak in the treatment of tacit knowledge as a crucial element of an organization's knowledge base and in the use of knowledge for strategic options. There seems to be quite a need for further conceptual and empirical work beyond the general and simple notion that more information for more people enhances the knowledge base of the organization.

7.4. Processes of Knowledge Generation

To understand the intraorganizational processes of knowledge generation, it is important to look back to the cultural and structural devices as well as the role of visions and leadership in the process. Several questions arise beyond those raised above: Whose knowledge is relevant for specific decision-making situations, problem solving, and the exploration of strategic options? How is relevance to be defined? How does one tap into knowledge that may exist among organizational members but that has not been used as a knowledge base for organizations because it was regarded as irrelevant or even as part of the private sphere of organizational members? How can members of the organization be encouraged to contribute knowledge perceived by them to be relevant for the organization? How can key decision-makers avoid becoming overburdened with information of little or no relevance to the decisions, problems, or strategic options under consideration?

Even more important and more difficult research focuses on the question of how to ensure the sharing of knowledge at an interorganizational level (Child, 2001; Lyles, 2001, Lane, 2001). How do strategic alliances allow for a transfer of relevant knowledge? What kind of role do international joint ventures play in this context? Are expatriates an important, but still underutilized, source of knowledge in the organization (Berthoin Antal and Böhling, 1998)? What kind of network device is appropriate to ensure undistorted and effective sharing of knowledge between organizations? How much cultural similarity is necessary for such sharing? Are there available structural devices that have proved productive in the exchange of knowledge between organizations? In this context, the role of boundary-spanning personnel, boards, and job rotation among organizations are again some of the structural and procedural devices to be examined from the point of view of interorganizational knowledge transfers. The more important mergers, acquisitions, and strategic alliances become, the more central these questions of ways and means are and the more relevant their effectiveness at transferring knowledge among organizations is. Quite important in this context is again the distinction between knowledge and information. What can be considered knowledge in one organization may just be information in another, for the two contexts may be completely different because of markets, cultures, and technologies that distinguish the organizations.

7.5. Creativity and Knowledge Generation

Research on knowledge management and the knowledge-based firm thus far has reflected little of the literature on individual creativity, group creativity, organizational creativity, and devices to improve the creativity of groups, teams and organizations. The first step to enhance current understanding of the knowledge-based organization and knowledge management is to link these two bodies of literature and explore the extent to which creativity research and the relevant methodologies in this field can contribute to the understanding of knowledge creation and utilization in an organization.

7.6. Knowledge and Organizational Change and Transformation

Within the last decade a large amount of literature in management has been written about organizational change and transformation. It stresses the need for companies to monitor the environment carefully and closely in order to understand the need for change as early as possible and even take steps to anticipate changes being wrought by markets, technologies, demographic developments, and sociopolitical and cultural discontinuities. Knowledge plays an important role in different forms of organizational learning. Single-loop learning can still draw heavily on existing knowledge, whereas double-loop learning and deutero-learning require greater depreciation of existing knowledge and the acquisition of new knowledge. One must ask, however, whether processes of organizational change are stimulating enough development of a new knowledge base for the future environment of the organization? How large does the knowledge base in the organization need to be in order to generate a sufficient number of convincing options and scenarios for the future environment and potential strategic responses to it? If the knowledge base is not large enough, what kind of devices (such as consultants, boundary-spanning personnel, and boards) are needed for development of relevant bases of knowledge about the future environment and potential strategic responses to it (Berthoin Antal, Dierkes, and Hähner 1994)? How relevant and productive in this context are procedural devices developed in change management or concepts such as scenario planning? In the process of rapid change, is a core knowledge base relevant to an organization? What does it look like under different contextual conditions? Is there a need for complete redefinition of the necessary knowledge base? How do organizations productively survive significantly different environments for long periods while dealing with these kinds of changes? Are there core elements of the knowledge base preserved over long periods as crucial elements of a company's vision or are those elements constantly changing most of what is in their organizational memories in order to survive as an organizational entity? Is it necessary in periods of profound environmental change to replace most members of the organization with new members, a new leadership, and a new vision, or can parts of the organization and culture remain intact during periods of rapid organizational change? What is the experience with the core knowledge of industries that traditionally deal with significant, but partially unpredictable, changes in their environments. For example, the fashion and entertainment industries have long experienced constant and often profound changes now regarded as characteristic in industries that have traditionally operated in more stable and predictable environments. These questions, too, link change, adaptation, and transformation with the study of the knowledge base of an organization.

7.7. Knowledge Management and the Future of the Organization

In management literature consensus is growing that the business corporation of the future will be a rather network-type of organization in which teams and projects are key elements of the organizational structure (Hedberg and Holmquist, 2001). Thus organizational structures will be more temporary in nature and more flexible than they presently are. Some organizations may even become virtual organizations. The links between the creation, use and management of knowledge and the loosely and temporarily defined organizational structures in and between organizations are still under explored and must be studied further. An important question is whose knowledge base (individuals, groups, communities of practice, knowledge communities, subcultures) can contribute knowledge relevant to the tasks at hand when project teams and other kinds of temporary organizations are formed for the purpose of solving problems, preparing decisions, or outlining strategic options?

More and more, the prerequisite for knowledge management in organizations is an effective institutional device structurally designed and grounded in culture and defined by whoever has the knowledge to effectively form these project teams and other forms of temporary organizations. Brokers and agents of knowledge communities or communities of practice in and between organizations are probably the most crucial individuals or organizational units with increasingly important functions in the organization of the future. Such brokers of knowledge are necessary in order to bring together and coordinate knowledge of the highest potential relevance for solving a given problem or preparing decisions or strategic options. A similar role can be witnessed in entire industries. An instructive example is the independent film industry, where relevant knowledge is embedded into different communities of practice and put together temporarily for specific tasks. Computer simulation experts specialized in crime stories are relevant for some modes of films. Others competent in the computer-aided simulation of historical events may bring a completely different, more appropriate knowledge base to other types of movies.

The key challenge is to decipher who knows what and to recognize whose knowledge base is the most relevant in a given context. The whole vision of more flexible, temporary, task-oriented organizations within the firm and within industries heavily depends on the effectiveness of those knowledge brokers. It also depends on the effectiveness of feedback information adding to the knowledge base of the brokerage function. In order to be effective, information about success or failure at putting together teams, competencies, and communities of practice for specific tasks within organizations or industries needs to be open and available to the knowledge brokers.

An interesting question is whether the role of the knowledge broker should be filled by staff members, by outside consultants, or by individuals specializing in this function in a particular industry. A wide field of research is developing in this context, and the examples I mention are only a first glimpse of what should be done.

7.8. The Study of the Knowledge-Based Organization: Methodology and Priorities

It is my hope that this sketch of research and ideas indicates one of the foremost responsibilities in this field of research, namely, the task of bringing together existing knowledge from different fields of inquiry, ranging from the study of organizational learning, creativity, governance, leadership, culture, and individual learning to the emerging field of research on knowledge management and the knowledge-based organization. What should be avoided is the reinvention of the wheel, which already seems to be taking place in some recent literature.

In terms of research strategy, the study of the knowledge-based organization would profit from time series and longitudinal studies, two methodological approaches rarely used in this field of inquiry. It is important to study the use of various knowledge bases in the organization over time, especially in periods of significant environmental changes. This approach is the only way to develop a better understanding of the relevance of unlearning existing knowledge and the underlying processes. It is also the only way to ascertain the extent to which visions held by organizations deeply committed to learning are stable, long-term elements of organizational culture. Also, case studies conducted in differing situations, markets, and historical periods and under differing technological conditions and sociopolitical environments would help improve the understanding of the specificities of the knowledge-based organization. In addition, comparative research is needed because different cultures may employ different strategies to improve their use of the various knowledge bases within the organization. Comparative research may thereby provide for learning from very different situations. Historical research, too, is critical in this context, specifically in those industries that have traditionally relied heavily upon knowledge, especially tacit knowledge.

All these endeavors would indicate whether knowledge is becoming a more important factor in our organizations than it has been in the past. If researchers are to improve their understanding of these developments, and if they are to help knowledge managers in organizations that are responding to opportunities and risks in rapidly changing and partially unpredictable environments, then knowledge about managing tacit and explicit knowledge has to be fully exploited. The faddish dimension of the current interest in the knowledge-based organization and knowledge management must then be distinguished from the core purpose of research and conceptual work in this field.

Chapter Three
CONTINUITIES AND RUPTURES IN KNOWLEDGE MANAGEMENT PRACTICES[1.]

Dominique Foray
CERI, OECD

1. INTRODUCTION

There is little in the extensive literature on knowledge management, or in the related new business practices, to indicate what is new. The many historical examples suggest that there have always been elaborate forms of knowledge management, whether in the context of the network of Cistercian abbeys that constituted an effective system for the dissemination of new agricultural techniques (Gimpel, 1995); the company craft guilds in the Middle Ages that took on the essential functions of embodying and transmitting practical knowledge (Epstein, 1998); or, finally, the large metallurgical and chemical companies in the early 20th century which ensured the links between research and development and learning through application by doing (Caron, 1997). Knowledge and learning have thus always been central to the economy and "learning organizations" have always existed. Why, then, talk about new developments and ruptures?[2.]

An initial answer, which belongs in the realm of macro-economics (Abramovitz and David, 1996, de la Mothe and Paquet, 2000), is akin to saying that knowledge-based economies reflect the fact that, first of all natural resource-based wealth provides a comparatively weaker explanation for the differences in growth rates and national wealth. Second, it supports that capital and work have more weight in explaining them, due to the changes in quality that take place when improvements in knowledge are reflected in workers and equipment, and the organizational transformations that follow from these improvements.

---

1. This text would not have seen the light of day without the insistence and support of Fred Gault. I also extend my thanks to Louise Earl, John de la Mothe, Kurt Larsen, Stephane Lhuillery, Keith Pavitt and Larry Prusak for their comments and suggestions.

2. The previous book, a product of the co-operation between PRIME and Statistics Canada, provides an excellent panorama of the ruptures and discontinuities at both the macro-economic and industrial organizational levels (de la Mothe and Paquet, 2000).

We will attempt to develop another order of response, one that falls within the purview of innovation strategies and organizational behavior. New industrial practices, grouped under the heading of "knowledge management", are emerging for the following reasons:

(1) First, some of the older practices which helped in knowledge management, "comme Monsieur Jourdain faisait de la prose" ["as Monsieur Jourdain talked in prose"] stated, no longer work;[3.]
(2) Second, some entirely new problems have emerged; and
(3) Third, our understanding of the phenomena pertaining to learning and the transmission of knowledge is increasing; this, in turn, provides an opportunity to forge new tools and new techniques of knowledge management.

2. SOME PRACTICES NO LONGER WORK

The memorization and transmission of tacit knowledge has always been ensured by internal institutions (the corporation craft guild, see Epstein, 1998, the internal labor market, see Lam, 1999) and external organizations (professional networks), in which this was an essential function. The unobservable element, in this case knowledge, was handled in a contingent manner in the context of employment and industrial relations policies which effectively ensured the memorization, circulation and transmission of knowledge. However, these institutions have largely disappeared or find themselves in profound crisis. For instance, in large French companies, a new engineer was normally hired a year before the old one retired in order to ensure that knowledge was passed on in the context of an extended master-student relationship. In such cases, the conditions were propitious for ensuring that the professional community itself ensured the memorization and transmission of knowledge from one generation to the next. However, the system is so costly that it is no longer used. These days, a young engineer arrives a few weeks before the old one passes him/her the reins. Naturally, the transmission of knowledge is only very partial. Thus, the old system for transmitting new knowledge management practices has to be replaced by a new one, which might, for instance, be based on a codification of knowledge that would enable a new arrival to use this written memory as a learning program (instruction manuals, maintenance documents, expert systems). Naturally, codification (which involves explaining tacit knowledge, expressing it and recording inscribing it in a certain format) mutilates knowledge. However, to borrow from the wise words of J. Goody (1977), "la recette écrite remplit partiellement le vide créé par l'absence de la grand-mère" ["the written recipe partially fills the void created by the grandmother's

---

3. I am borrowing this analogy from Larry Prusak, who developed it during an oral presentation at the Ottawa Seminar.

absence"]. An economist would say that the marginal cost of reproducing knowledge, in the absence of an experienced engineer (who has gone), will fall as a function of the codified knowledge that is at the disposal of the young engineer.

Social institutions, the standards and social norms which facilitated the sharing of knowledge (a form of social capital) are in crisis. This implies a need for new types of incentives and the reconstruction of a logic of rationale for knowledge sharing to replace the one that has disappeared. The new forms of knowledge management essentially involve methods of knowledge sharing and circulation.

Other practices no longer work. The principle of lifelong careers and very long-term attachment to the company led to a kind of common destiny between the employee and his/her company. From that point on, the individual's knowledge was an almost indetachable part of the company's intellectual heritage. Here again, recent developments in terms of turnover, mobility and flexibility make it necessary to invent new forms of knowledge retention - again, through either codification or the implementation of strong legal mechanisms to protect the company's intellectual heritage, or through human resources policies that are better suited to maintaining skills.

In all, recent knowledge management methods attempt to create new modalities for memorization and transfer (organizational memory) in order to encourage people to share their knowledge and reduce the loss of knowledge resulting from the high mobility rates among staff. These methods appear to be the expression of a crisis of communal action the company as a collective good (Le Masson, 2000). Communities no longer seem capable of "spontaneously" taking charge of the essential functions of knowledge memorization, transfer and sharing.

3. NEW PROBLEMS

Relatively new problems have forced the introduction of explicit forms of knowledge management. These involve the increasingly central role of innovation as a condition of business survival and, consequently, the growing importance of what we will refer to as the "capacity for innovation". These also involve a cluster of structural changes which can be grouped, for the sake of convenience, under the heading of "new economy". These changes concern the extension of the role of market transactions in the field of scientific and technological knowledge, the massive use of information and communication technologies and the importance attached to "intangible" resources in the stock market valuation of companies. All of these changes require that companies invest in extensive formalization and systematization of procedures for the identification, storage and evaluation of intangible resources.

3.1. Managing Knowledge In Order To Increase The Capacity For Innovation

Most recently, there has been a marked increase in competition, linked to market liberalization, globalization, the use of new information technologies which increase the transparency of markets and, finally the domination of the "winner-take-all" principle in some markets (Shapiro and Varian, 1998). This trend has forced companies to adopt much more aggressive innovation policies. Innovation is one of the ways of "s'extraire de la concurrence" ["getting out of the competition"], which is to say situations in which activities are not very profitable.[4.] Innovation provides an opportunity to set oneself apart and to temporarily assume a monopoly position; this provides an opportunity to make a profit either by increasing prices (product innovation and above price competition technological competitiveness) or by reducing costs (process innovation and active price competition competitiveness). Innovation then becomes a condition of wealth, if not survival. The cost of missing the boat on an innovation (bypassing and ignoring a "good idea") becomes enormous. We no longer have the luxury of missing out on one or for two innovations. Thus, it becomes essential to introduce planned strategies for the collection and documentation of ideas and suggestions by employees. In addition to this type of knowledge management, processes for stimulating creativity become essential.

As soon as innovation takes a central place in a business' strategy, its internal and external resources have to be controlled. This is the direction of the new intra- and inter-organizational learning systems, as well as of the development of efficient information research procedures.

First, the new intra-organizational learning systems are characterized by a wider distribution of the knowledge production function (Steinmueller, 2000a). The previous period, referred to as that of mass production, was characterized by tremendous specialization in knowledge production or reproduction functions within the company and between sectors. Under the new systems, knowledge production is more inclined to be distributed throughout the community. This translates into the extension of experimental forms of learning in companies, which call on employees to conduct experiments in the course of their everyday production activities in order to select the best strategy for the future. This means that more and more economic agents are producing knowledge even though this is not the main reason for their activity. One of the spectacular forms of experimental learning is the one in which users are involved in the innovation process. As users of the new technologies, users get to become the developers and, hence,

---

[4.] Naturally, there are other ways of getting out of the competition. Some are more negative (using market imperfections in terms of transparency and access to information and resources), while others are more positive (creation of loyalty through good brand name policies or other ways).

an essential source of knowledge (von Hippel, 1988). In some areas (such as open source software), this trend can lead to "users-only innovation systems" (Lakhani and von Hippel, 2000). This trend, which makes it easier to develop the productivity potential of professional communities, also imposes significant constraints: first, experimental learning - that which is done on-site, outside the protective confines of the research laboratory - may conflict with normally expected performance. Thus, it is important to introduce mechanisms that will enable organizations to tolerate a certain amount of failure, for instance by preventing the effects of these failures from spreading throughout the system. Second, significant economic problems arise, such as the allocation of costs and revenues to activities that are subject to shared economies. Third, it is not enough to let people experiment. It is also necessary to establish mechanisms that will encourage these people to evaluate, document and share their experiences. In this sense, the emergence of new learning systems in which knowledge production is more collectively distributed is inseparable from the management of knowledge.

Second, the transformations that are currently under way also concern inter-organizational learning through increased co-production of skills (for instance, between a supplier and a producer). Such co-production - which may take various forms, from establishing a strong relationship between a supplier and a user to creating a complex set of co-producers based on the modularity of the product - significantly increases the requirements in terms of knowledge management. Thus, knowledge produced at different sites is most often "sticky" (von Hippel, 1994), which is to say that it is not readily transferable from one site to another. Therefore, it is very difficult to consolidate all of this technological knowledge in one place for the purpose of resolving all the problems that arise. The problem-solving activity has to be transferred from one place to another through a process of successive repetitions iterations. This can work properly only if there is an inter-organizational interface between the different teams. Another approach is to invest in the codification of knowledge (i.e. 'un-sticking' information) in order to transfer the knowledge instead of the activity.

Third, the tilting of our economies in a "Simonian" world in which it is no longer information and knowledge which are scarce, but rather attention, also forces companies to develop specific skills for managing attention, filtering and discrimination of information (Simon, 1982). This world, in which attention has become a scarce resources (or, to use L. Prusac's terms, time has become a scarce resource), may be characterized by a very low probability of finding useful information (or knowledge), even though it has to exist somewhere.

The wealth of information combines with the increasing dispersion of knowledge (linked to the countless number of local and specific sites where information is produced) and to the increase in the division of knowledge (associated with the increasing division of labor in the production of knowledge and, thus, to specialization in various fields) to create a huge stockpile through which it is very hard to maneuver. The difference between the probability that the information sought exists somewhere and the probability of finding this information is increasing. New skills, cognitive capacities and technological tools are called for and their acquisition is an integral part of knowledge management programs.

Thus, the extension of the capacity for innovation essentially implies the management of knowledge, both at the level of internal learning and at the level of the co-ordination of external resources.

3.2. Managing Knowledge In Order To Enter Into The New Economy

The extension of knowledge markets, the dissemination of information technologies and new methods for the evaluation of intangible assets are three characteristics of the new economy which require the introduction of explicit knowledge management methods.

First, never before has there been such growth in market transactions in connection with knowledge. The increase in the rate of patent applications, the impressive growth in revenues arising from the granting of licenses and the explosion in costs associated with intellectual property settlements are all indicators of the current development of the "knowledge-based market economy" (Arora, et al., 2000). Yet, knowledge markets are, by definition, inefficient markets (Teece, 1998). Buyers and sellers are not well informed about the commercial opportunities (no one knows who has what or who wants what). There are problems associated with revealing the characteristics of the product. Intellectual property rights, even though they can reduce the first two difficulties, are fragile, uncertain and heterogeneous. The product (or consumption) unit is not clear. Knowledge is sold neither by weight nor by size! Finally, the characteristic of this product that is hard to control generates massive externalities which continuously sap the foundations of the market transaction, which assumes that the product can be owned. At this point, knowledge management can be interpreted as an effort to create less efficient market conditions. From this point of view, intellectual property policies clearly form part of knowledge management. The issue is not only one of protecting innovations by applying for patents, even though this is a central element which, in itself, presents enormous problems. Rather, the issue is one of ensuring preventive management, which is to say checking to be certain that the research and innovation areas that are being targeted are free. As well, intellectual property also concerns trade secrets and legally protected codified know-how (often called proprietary information), such as technical drawings and training, maintenance and operating manuals. It is difficult to manage this aspect of intellectual property since this information has often not been collected or consolidated, and it remains poorly identified within the company. It then becomes evident that an effective intellectual property strategy involves codifying and organizing the company's knowledge. This goes well beyond the procedures and products that will be patented. Finally, a good intellectual property policy provides an opportunity for effective management of externalities through cross licensing (Grindley and Teece, 1997).

Second, before they can become solutions, the new information and communication technologies present as problems! The paradox of productivity can be expressed very simply as the delay between the appearance of new knowledge tools and instruments and the persistence of old forms of organization. Decentralization of the processes of knowledge creation, which are made possible by the new information technologies, requires the development of inter-organizational interfaces in order to minimize the time required to establish and carry out a transaction. It then becomes a matter of moving to a higher level of systematizing organizational skills and procedures. The management of knowledge, particularly in terms of the codification of procedures, is central to these changes (Steinmueller, 2000b).

Third, the evaluation of intellectual capital becomes a decisive element in the evaluation of the company. It appears that the stock market valuation of a company increasingly depends on the value of its intangibles. Here again, the management of knowledge involves techniques for the identification and quantification of intangibles in terms of the company's knowledge base.

3.3. Managing Knowledge Means Negotiating Between Conflicting Requirements

The requirements associated with intellectual property may conflict with those arising from the company's need to be open and to connect with external networks and sources of knowledge. Thus, the company is forced to negotiate and make compromises between openness and secrecy. Managing knowledge increases the risk of losing control over its intellectual capital. One of the benefits of a situation in which knowledge is not managed is that it remains shielded, invisible to others and, therefore, very difficult to imitate or reproduce. Defining knowledge, codifying it, providing incentives to encourage employees to describe and disseminate their skills are all high-risk activities from the point of view of the control that a business would like to exercise over its intellectual capital. Therefore, knowledge management involves accepting compromises and negotiating full control (so to speak) of the knowledge produced within the company. This was perfectly illustrated by Diana Hicks (1995) with her explanation of how some companies publish (freely provide knowledge) to broadcast their skills and attract scientific partners.

4. TOWARDS AN INCREASED UNDERSTANDING OF THE WORLD OF THE INTANGIBLE!

The management of knowledge, as an activity, requires project engineering in the form of tried and true tools and techniques which have themselves been built on the basis of general economic progress and knowledge advances in the economics and management of knowledge, as a discipline. Yet, since the work of Nonaka, Teece, von Hippel and many others, there has been significant progress in these disciplines, which has provided an opportunity to better understand the field and, thereby, the possibility of new tools. We will limit ourselves to giving two examples of this improvement in our understanding of the phenomena, which may open the door to improved knowledge management practices.

The works of von Hippel and associates are fine examples of the advances in our understanding of the learning process. The development of a "situated perspective" highlights the importance of the physical context of learning. This context, which is often ignored in analysis, is in fact an essential component in the process. This perspective is based on the pragmatic argument that knowledge is not absolute, but must be defined in relation to a specific physical context. From a concrete point of view, Tyre and von Hippel (1995, 1997) review the reasons why an engineer will pay frequent visits to a user in order to settle a technical problem. The key reasons touch on the importance of the visual experience in understanding a situation (I "see" the problem, whereas the user does not), the importance of the context within which the machine is used, and the interaction between the user and the machine. In other words, solving a problem involves more than simply choosing a good representation of the problem and selecting a solution from an inventory (Simon, 1982). Rather, it is a matter of drawing from the physical context in which the problem arose. Such an understanding of the situational nature of learning provides an opportunity to design principles of location and "optimal mobility" for experts as a function of the operational stages. In particular, it becomes evident that mobility must also concern the engineering consultant who has to visit the user and the doctor who has to make a house call in order to benefit from this essential component of learning, which manifests itself in the physical context.

Another example of progress made in our level of understanding is provided by the work of Hansen (1999), who presents the problem of negotiating between the requirements of searching for information and transferring knowledge. In a large organization, a team may be faced with the problem of searching for information. From this point of view, there would be an advantage in building a system of weak ties (distant and infrequent connections). Maintaining a weak tie is inexpensive, which makes it possible to "keep an eye" on the entire organization. In addition, weak ties reduce the risk of redundancy in the collection of information. Transferring knowledge presents a different problem. It requires that strong ties be built, particularly when the knowledge is tacit and specific. The differentiation between information search and knowledge transfer issues provides an opportunity to reconcile two types of literature which provide initially contradictory solutions (weak tie, strong tie). Therefore, these works often provide an opportunity for fine management of these ties and, thereby, for successive improvements in information search and knowledge transfer.

Just as progress in scientific instrumentation makes in it possible to observe phenomena that were previously invisible, progress in the "sciences de l'innovation" ["innovation sciences"] introduces a world that had previously been ignored. The exploration of this universe makes it possible to improve our understanding of the process of knowledge production and use and, in the end, provides new operational opportunities.

5. BEYOND THE MANAGERIAL AND ECONOMIC APPROACHES ...

There is a rupture and discontinuity in the current knowledge management practices because some of the older practices, buried in human resources and employment policies, no longer work. For this reason, it becomes important to develop explicit and deliberate procedures for managing knowledge, which are in a way detached from the employment policies and are connected to the institutional management of knowledge.

There is a rupture and discontinuity because new problems arise, problems that are largely linked to the urgent need to master innovation and control the phenomena of the new economy. It then becomes important to implement explicit forms of knowledge management in innovation.

We believe in a certain degree of "scientific push" in the management and innovation sciences. As our understanding of the world of the intangible improves, tools and operating methods are introduced, tested and improved, which encourages practitioners to develop knowledge management methods on the basis of scientific methods.

However, beyond this economic and managerial line of argumentation, we should perhaps review the trends that have been studied until this point from another perspective, that of political sociology.

A systematic review of the management literature from the 1960s and 1990s led Boltanski and Chiapello (1999) to suggest the emergence of a "new age of capitalism". Their thesis was sufficiently developed and convincing to lead to an attempt to look at the new knowledge management practices as an expression of this new spirit.

The thesis presented by these authors is the following. In order to maintain its power to motivate, capitalism must, in each period, provide those who participate in the process of capitalist accumulation (specifically for senior managers and engineers) reasons to get excited (including those who are not the primary beneficiaries) and reassured (security aspect) and, finally, justify, in terms of the common good, their own co-operation in the face of accusations of injustice and inequality. Yet, the "artistic and social" critics whose criticisms which acquire new strength and forms from one period to the next, regularly undermine the system with the reasons and justifications presented.[5.] The power of motivation behind the spirit of capitalism is then questioned and it becomes necessary to create a new system of reasons and justifications. Thus, the succession of "spirits of capitalism" follows along with the implementation of successive new systems of argumentation and representation, which are capable of renewing the grounds for motivation for those who participate in the capitalist enterprise "L'enjeu sera de rendre les nouvelles formes d'accumulation séduisantes à leurs yeux (la dimension excitante de tout esprit), tout en tenant compte de leur besoin de s'autojustifier (en prenant appui sur la référence à un bien commun) et en édifiant des défenses contre ce qu'ils perçoivent, dans les nouveaux dispositifs capitalistes, comme menaçant pour la survie de leur identité sociale (la dimension sécuritaire)" (p. 60) ["The challenge will lie in making the new forms of accumulation attractive to them (the element of excitement in any mind), while taking into account their need to justify themselves (by relying on the reference to a common good) and by building defenses against that which they see, in the new capitalist mechanisms, as a threat to the survival of their social identity (the security aspect)"]. The 1990s can then be seen as the beginning of a new age, based on a new representation of the company and economic processes - better equipped to respond, under the current conditions, to the artistic and social criticisms. Who sees nothing but the managerial point of view on the importance of knowledge held by individuals, on the fact that everyone in the company is a source of knowledge and that sharing and circulating knowledge becomes a priority of the organization, that this discussion forms a fundamental contribution to this new aspect, which is designed to provide new reasons for being interested in the capitalist enterprise, as well as new arguments for justifying its excesses?

---

5. The artistic criticism reproaches capitalism for losing the sense of beauty and grandeur through mass standardization and marketing, affecting not only daily objects, but also works of art and human beings. As for the social criticism, which is inspired by socialists and Marxists, it reproaches capitalism for the selfishness of individual interests and the growing misery of the common people. Naturally, this two-fold message has taken different forms over the centuries.

PART III

MEASUREMENT

Chapter Four
CREATIVITY, INNOVATION AND BUSINESS PRACTICES IN THE MATTER OF KNOWLEDGE MANAGEMENT

Réjean Landry
CHSRF/CIHR Chair on Dissemination and Uptake of Research
Département de science politique
Université Laval

and

Nabil Amara
Département de science politique
Université Laval

1. INTRODUCTION

This chapter addresses one question: do business practices in the matter of knowledge management determine creativity of manufacturing firms? Creativity is measured with a question concerning the extent to which product and process innovations are developed primarily by the firm, primarily by other organizations, or jointly by the firm and other organizations. To deal with this question, we review the literature on innovation in order to see why creativity and knowledge came to be added to the traditional explanatory variables of innovation. The literature suggests distinguishing codified from uncodified business practices of knowledge management. This chapter distinguishes five forms of uncodified practices of knowledge management: business network, information network, research network, participation, and relational assets. As for the forms of codified practices of knowledge management, we distinguish four forms: acquisition of knowledge embodied in advanced technologies, creation of knowledge through R&D activities, capacity of knowledge management through the ratio of scientists and engineers in firms, and structured mechanisms for sourcing knowledge. Barriers concerning the lack of cooperation with providers of knowledge, the number of employees and sales are also used as explanatory variables of the creativity of firms. The data used for this chapter come from an innovation survey administered from April to June 2000 to 440 manufacturing firms of diverse industries in Montérégie, a region in the South-West of Montréal. These data are used to estimate two multinomial logit models using creativity as its qualitative dependent variable. The qualitative dependent variable is defined as follows: innovation developed primarily by the firm, innovation developed jointly by the firm and other organizations, and no innovation. A model is estimated for product innovations and another for process innovations.

This chapter addresses one question: Do business practices in 1the matter of knowledge management determine creativity of manufacturing firms? In this chapter, knowledge management is defined as a process to foster knowledge for developing innovation. More specifically, knowledge management focuses on:

(1) The creation of knowledge;
(2) The importing of knowledge through uncodified business practices;
(4) The importing of knowledge through codified business practices; and
(4) The translation of knowledge into innovation.

Although there is an expanding body of studies on knowledge management and innovation, these studies tend to suffer from three deficiencies: first, although the conceptual studies are expanding at a very rapid pace, empirical studies attempting to test the impact of knowledge management on innovation are to our knowledge, very scanty. The second deficiency derives from the specification of the dependent variable innovation which, in most studies, models innovation as present or absent without taking into account the degree of novelty of innovation. The third deficiency derives from problems associated with the independent variables considered in innovation studies which have not yet accumulated much significant empirical evidence on the various dimensions of knowledge. Part of these deficiencies result from the fact that although knowledge is now considered as one of the major determinants of innovation, there is not yet a general theory of innovation suggesting explicit linkages between diverse dimensions of knowledge and degrees of novelty of innovation.

Although we assume that knowledge is one of the most powerful drivers of innovation, we know very little about the business practices of knowledge management that induce manufacturing firms to innovate. The purpose and contribution of this chapter are to build and extend from the existing body of knowledge: first, by using a specification of innovation that distinguishes innovative firms according to degrees of creativity; second, by using a specification of business practices in the matter of knowledge management that distinguishes uncodified from codified practices; third, by testing the impact of business practices in the matter of knowledge management on innovation.

The chapter is organized in five sections. The first covers the conceptual issues regarding the specification of the dependent variable innovation and independent variables regarding business practices in the matter of knowledge management. The second part of the chapter presents the survey data used to develop a model that is introduced in the third part of the chapter. Then, the next part of the chapter presents the results of multinomial logit models. The last part of the chapter discusses the findings and derives implications for future research and public policies.

2. CONCEPTUAL ISSUES

Empirical studies on innovation and knowledge management tend to suffer from two conceptual problems: problems related to the specification of the dependent variable innovation, and problems associated with the independent variables considered. Let us initially consider the first problem.

2.1. Specification Of The Dependent Variable Innovation

Until the Oslo Manual (1996b), innovation was measured indirectly with various indicators, notably patents indicators and R&D indicators. The guidelines proposed in the Oslo Manual for collecting technological innovation data introduced and contributed to disseminate the use of a direct measure of innovation through two questions like: "During the last three years, did your firm offer new or significantly improved products (goods or services) to your clients? Yes or No", and "During the last three years, did your firm introduce new or significantly improved production/manufacturing processes? Yes or No". These questions represented a genuine contribution to the improvement of the measurement of innovation by showing that a significant proportion of innovations did not require R&D investments or patents. Hence, for small and medium enterprises, one finds that roughly twice as many firms develop innovations than there are investing in formal R&D activities. For instance, the Eastern Townships regional innovation survey that we conducted in 1998 indicated that 22 percent of the manufacturing firms obtained R&D tax incentives, 44 percent claimed that they were investing in formal R&D activities and 87 percent of the respondents indicated that they had either offered new or significantly improved products or introduced new or significantly improved production/manufacturing processes (Landry, 1998). Regional innovation surveys conducted in Québec by our team (Landry et al, 1995; Landry et Fréchette, 1998; Landry, Lamari and Amara, 2000; Landry, Amara and Lamari, 2000) indicated that the proportion of firms introducing innovations has tended to increase rapidly since the initial formulation of the guidelines proposed in the Oslo Manual and, furthermore, that the proportion of innovative firms tends to vary drastically from region to region. Our contention is that the current measure of innovation (offering new or improved products or introduction of new or improved processes) is becoming less and less useful for the scholarly explanation of the innovation process and as a policy tool, as the proportion of innovative firms approach 100 percent.

Indeed, as the proportion of innovative firms increase much over 50 percent, it becomes important to ask questions about the degree of novelty of the products or processes considered as innovations. This scenario had been envisaged in the guidelines proposed in the Oslo Manual. The conceptualization of novelty took five complementary directions: time, resources, frequency, content and sources. Let us discuss briefly these conceptualizations in turn. One very simple way to conceptualize the degree of novelty of innovations is to ask firms to indicate how many months elapsed between their first investments in the development of their innovations and their first introduction in the firm or their first offer on the market. The assumption behind this question is that the degree of novelty of innovation increases as the number of months required for developing innovations increases. Landry, Amara and Lamari (2000b) have used this conceptualization of the degree of novelty of innovation with the justification that the answers to this question are easy to provide by respondents and constitute facts, not perceptions.

As for the second conceptualization where the degree of novelty is measured in terms of financial investments in the development of products or processes (in terms of investments/sales), we have found that this particular figure is correlated to volumes of sales, thus indicating that small firms tend to invest relatively more in innovation than large firms do. This has led us to think that the ratio of investments in innovation is not a valid measure of the degree of novelty of innovation.

Another measure of the degree of novelty of innovation was used by Statistics Canada in its 1999 Innovation Survey. It is based on the following question: "Was this innovation: a) a world first? If not, was it a Canadian first? A local market first? Yes, No, do not know". In our 1995 regional innovation survey of Chaudière-Appalaches (Landry et al., 1995), we found that world first was highly but inversely correlated to the size of firms: the smaller the firms and the lower their percentage of sales shipped outside Canada, the higher their likelihood to answer that their innovations were world first introductions. Our contention is that for small firms, this particular question does not measure facts, but perceptions or ignorance. However, we have also observed that the answers provided by the medium and large firms were less counterintuitive because world first introductions of innovations were correlated with high percentages of exports.

In innovation surveys, frequency with which innovation is introduced is measured with a question like the following: "Please indicate how many new or significantly improved products were offered in the last three years". The answer to this question provides evidence on the culture of firms by showing to what extent the development or improvement of products or processes constitutes either a regular recurrent pattern of behavior or an exceptionally rare pattern of behavior. Therefore, this measure does not provide evidence regarding the degree of novelty of innovations.

Finally, and more recently, a new measure of novelty of innovation was developed by Eurostats for the second Community Innovation Survey (March, 1997). The proposed question is asked immediately after the question on development of products or processes and it is phrased as follows: "... who developed these products (or processes)? Mainly other enterprises or institutes? Your enterprise and other enterprises or institutes? Mainly your enterprise?" To our knowledge, this question has not been used to conceptualize novelty of innovation, but more likely the internal capability of firms regarding the development of innovations. In this chapter, we propose to use this question as an indicator of the creativity of innovative firms and to distinguish firms according to four categories of creativity:

(1) Innovations developed mainly by the firm interviewed;
(2) Innovations developed jointly by the firm and other firms or institutions;
(3) Innovations developed mainly by other firms or institutions; and
(4) No innovation developed by the firm interviewed.

2.2. Problems Associated With The Independent Variables

In spite of a large body of empirical literature on the determinants of innovation, there is not yet a consensus regarding the categories of factors that explain innovation. The pioneering studies on innovation implicitly assumed that innovation was caused by events that were either internal to firms or by external events crystallized into the forms of patents. The emerging theories of innovation lay the stress of four categories of ideas:

(1) Innovation is not only caused by discrete events, but also by problem-solving processes (Dosi, 1982);
(2) Innovation is not only caused by factors internal to firms, but also by interactive processes involving relationships between firms with the different actors of their environment (Kline and Rosenberg, 1986);
(3) Innovation is not only caused by an additional productivity of operations, but also by a systemic productivity of relations (Foray, 1998, 2000; Lengrand and Chatrie, 1999) generating diversified learning processes: learning-by-using, learning-by-doing, learning-by-sharing (Rosenberg, 1982; Lundvall 1988, 1995;

Malerba, 1992). Learning may arise from internal or from external sources of knowledge (Dogson, 1991). External learning refers to the absorption capacity of firms (Cohen and Levinthal, 1990); and

(4) Innovation is not only caused by isolated learning, but also by social processes of exchange of knowledge generating a system, an innovative system (Johnson, 1995; Lundvall, 1992; Acs, 2000; Braczyk, Cooke, and Heidenreich, 1999; Cooke et al, 2000; de la Mothe and Paquet, 1998; Edquist, 1997; Edquist and Hommen, 1999; Holbrook and Wolfe, 2000; Landry and Amara, 1998; Niosi, 1993), a "système social d'innovation" (Amable, Barré and Boyer, 1997; De la Mothe, 2000), a "milieu innovateur" (Maillat, 1995; Storper, 1997) or an innovation cluster (Porter, 1999; 2000).

These buoyant ideas are leading to revisions in innovation theories. Traditional theoretical and empirical studies focused on explanatory factors internal to firms such as investments in R&D and production and acquisition of patents. New approaches are now developed to include knowledge management as explanatory factors of innovation. The growing importance of knowledge as a production factor and as a determinant of innovation can be explained by the continuous expansion of the amount of technical knowledge accumulating over time, and by the use of communication technologies that makes the knowledge available very rapidly on a worldwide scale. (Arundel et al, 1998; Cowan and Foray, 1998).

Compared to the traditional theories of innovation, the recent theories of innovation lay much emphasis on the strategic importance of relational tools rather than only on technical tools, and on knowledge networks rather than only on technological networks. The development of knowledge-based innovation requires the capacity to implement technical and relational tools (Lengrand and Chatrie, 1999): technical tools refer to the acquisition and utilization of new information and communication technologies. These technical tools do not create competitive advantages because they are readily available to others. The creation of competitive advantages rests on relational tools: that is the way of doing business, both in the internal and external environments of firms. As for networks of cooperation, Lengrand and Chatrie (1999) claim that knowledge networks appear as a new form of cooperation networks, taking multiple forms that are added over and above the technological networks that they define as the first generation of cooperation networks.

The evolution from a technological network perspective of innovation to a social network perspective of innovation has been led by the more and more pressing challenge to transform information into knowledge, i.e., information contextually connected to the development or improvement of products or manufacturing processes. Knowledge-based innovation requires not one but many kinds of knowledge. Furthermore, it requires the convergence of many kinds of knowledge held by different categories of actors.

2.3. Knowledge Management And Innovation

In the literature on knowledge management, the concept of knowledge management tends to be concerned by how firms measure and manage their assets of knowledge (Sveiby and Lloyd, 1987; Myeers, 1996; Prusak, 1997; Ruggles, 1997; Stewart, 1997; Wiig, 1997; Boisot, 1998; Davenport and Prusak, 1998; OECD, 2000a; OECD, 2000c, Von Krogh, Ichijo and Nonaka, 2000). In this chapter, knowledge management is defined in a more limited manner as a process to foster knowledge for developing innovation. More specifically, we consider that knowledge management focuses on:

(1) The creation of knowledge through codified business practices;
(2) The importing of knowledge through uncodified business practices;
(4) The importing of knowledge through codified business practices; and
(4) The translation of knowledge into innovation.

Our contention is that the capacity of firms to use knowledge for developing innovation does not lie only in their possession of knowledge, but also in the way the firms are structured to manage knowledge.

The literature suggests implicitly distinguishing codified from uncodified business practices of knowledge management. This chapter distinguishes five forms of uncodified practices of importing knowledge management: business network, information network, research network, participation, and relational assets. These uncodified practices of knowledge management are supported by social capital. As for the forms of codified practices of creating and importing of knowledge management, we distinguish four forms: acquisition of knowledge embodied in advanced technologies, creation of knowledge through R&D activities, capacity of knowledge management through the ratio of scientists and engineers in firms, and structured mechanisms for sourcing knowledge. These codified practices of knowledge management are supported by institutionalized mechanisms. Barriers concerning the lack of cooperation with providers of knowledge, the number of employees and sales will also be used as explanatory variables of the creativity of firms. Our overall hypothesis is that to be creative, innovative firms have to create and import knowledge through codified and uncodified practices of knowledge management.

Creation of knowledge through codified business practices: the percentage of sales dedicated to R&D can be considered as an indicator of in-house creation of knowledge by firms. Furthermore, it can be considered as a codified business practice of knowledge creation for these firms that succeed in obtaining R&D tax credits for their R&D activities. Indeed, the firms that obtain tax incentives have to codify their R&D activities. Small and medium firms that do not codify their R&D activities to the satisfaction of the officers of the Department of Revenue Canada do not get the R&D tax credits.

Import of knowledge through codified business practices I: the number of different advanced technologies used by firms can be considered as an indicator of the variety of knowledge embodied in equipment. Such imports of knowledge are codified business practices because they are made explicit in investment plans of firms, and because they are influenced by the explicit rules of the tax incentives regarding depreciation.

Import of knowledge through codified business practices II: the degree of institutionalization of the mechanisms used by firms to acquire knowledge usable for developing or improving products or processes is another indicator of the import of knowledge through codified business practices. Knowledge from external sources can be acquired through a multiplicity of codified mechanisms, ranging, as proposed by Leonard (1998), from observation, to non exclusive licensing, R&D contracts, equity/educational acquisitions, codevelopment, exclusive licensing, joint ventures, and finally to acquisitions/mergers. According to Leonard (1998:154): "... technology observation, nonexclusive licenses, and R&D contracts generally cannot be expected to provide new capabilities; they provide windows onto the technology and perhaps samples of its potential. Equity in a venture or 'educational' acquisitions provide an opportunity to investigate a technology in depth. Codevelopment inevitably results in knowledge 'bleedthroughs' from one company to another and thereby potentially launches a new capability. However, acquisition of a new core capability is likely only through joint ventures, company purchases, or mergers, which potentially transfer a technology in its full business context, with all four dimensions (window, access, new product/process, new capability) of the capability represented."

Import of knowledge through uncodified business practices: social capital is also used by firms to acquire knowledge useful for developing or improving products or processes. Social capital contributes to reduce the search, analysis and exchange of knowledge. Likewise, social capital enhances the exchange of tacit knowledge and reliability of the knowledge exchanged (Maskell, 1999). Social capital can take different forms, primarily trust, norms, and networks (Dasgupta and Serageldin, 2000; Fountain, 1998; Lesser, 2000; Putnam, 1993). Trust is developed over time through repeated series of interactions. Firms in milieu characterized by high levels of trust are more likely to innovate (Knack and Keefer, 1997, p. 1252). Norms of appropriate behavior also develop over time as a result of a series of interactions and exchange of knowledge. The norm that is the most often mentioned in the literature on social capital is reputation for trust-worthiness. Finally, networks develop as actors develop reliable and effective communication channels across organizational boundaries (Le Bas, Picard and Suchecki, 1998).

In this study, we use indicators derived from a regional innovation survey to measure diverse structural forms of social capital (Uphoff, 2000). The structural forms of social capital include roles, rules, procedures and networks that facilitate coordinating efforts, creating expectations, and lowering transaction costs. In this study, we measure three such forms of structural social capital: network assets, relational assets and participation assets. Although our questionnaire did have a question regarding trust, we have not included this variable in the present study because the index of trust was found to be unreliable (see Landry, Amara an Lamari, 2000 for details).

The questions used to assess the levels of these forms of social capital are:

Network assets: "Which of the following played a role as sources of information needed for suggesting or contributing to the development of new or improved products or manufacturing processes during the last three years? Please check all that apply in indicating the degree of importance on a scale ranging from 1 to 5 where 1= not important at all to 5= extremely important". We distinguished three types of networks: business, information and research networks. The business networks include clients, suppliers, competitors, suppliers of equipment, material and components, consultancy firms and related firms in your corporate group. The information network included trade fairs and exhibitions, professional conferences, meetings and publications, Internet or computer-based information networks, patent documents. As for the research networks, they included government research laboratories, technology transfer organizations, universities and community colleges.

The levels of the relational assets are assessed from responses to a question about the degree of agreement regarding the degree of acquaintance respondents had with four categories of economic actors. The question was: "Please indicate your degree of agreement with the following statements: I personally know managers and professionals involved in government agencies involved in 1) the promotion of economic development at the regional level; 2) the promotion of economic development at the provincial or national level; 3) university researchers working on projects related to my products; 4) my clients and suppliers. The levels of agreement were measured on a scale ranging from 1 to 5 where 1= do not know at all and 5= know extremely well."

The levels of participation assets are assessed from responses to a question about the frequency of participation of the respondents in meetings, associations and networks of manufacturing firms at four levels: regional, provincial, national and international. The levels of frequency are measured on a scale ranging from 1 to 5 where 1= never participate to 5=participate very frequently.

Capacity of knowledge management of firms: the percentage of scientists and engineers in the labor force of firms can be considered as an indicator of their capacity to create, acquire and translate knowledge for developing or improving products or processes.

Barriers to knowledge exchange: the creation, import and translation of knowledge in new or improved products or manufacturing processes can be hampered by the lack of cooperation with other firms, with government research organizations, and universities and technical colleges, as well as by the lack of information about technologies pertinent for developing or improving products or processes.

3. DATA AND DESCRIPTIVE STATISTICS

The data used in this study have been collected by a survey firm who conducted computer-assisted telephone interviews from April 10 to June 07, 2000. The questionnaire was adapted from the second edition of the Oslo Manual and included questions specifically defined by the sponsor of the survey: Valo Tech inc, a non-profit organization involved in the promotion of innovation in the Montérégie region. The purpose of the survey was to make a diagnostic regarding levels of innovation in the Montérégie region, a region located in the South-West of Montréal. The list of firms used to conduct the survey was provided by the Centre de recherches industrielles du Québec. The initial list included 2321 firms. After having excluded firms that were not manufacturing firms, no longer in operation, impossible to reach after more than 20 calls, or respondents out of town for long periods of time, the actual population of manufacturing firms was 1319. Out of this effective population, 817 firms refused to answer the survey (61.9 percent), 62 questionnaires were not completed (4.7 percent) and 440 questionnaires were completed and usable for a return rate of 33.3 percent. This return rate is higher than most surveys administered to manufacturing firms. Furthermore, a regional survey is especially appropriate to study the factors explaining the influence of social capital on innovation.

Overall, 68.5 percent of the firms have indicated they have developed or improved innovations of products or manufacturing processes during the last three years. The descriptive statistics of the variables used in this study are reported in Table 1. Overall, the average respondent has dedicated 6.3 percent of his sales to R&D activities, he has used 4.67 different advanced technologies in his manufacturing processes, he had sales amounting to 6.7 millions of Canadian dollars in 1999, he has exported 24 percent of his production outside Canada, he has 49 employees, he has ranked at 9.02 out of a possible maximum of 25 in the matter of business network assets, he has ranked at 6.51 out of a possible maximum of 22 in the matter of information network assets, he has ranked at 4.1 out of a possible maximum of 19 in the matter of research network assets, he has ranked at 9.59 out of a possible maximum of 19 in the matter of relational assets, and finally, he has ranked at 5.23 out of a possible maximum of 10 in the matter of trust assets.

4. MODEL

4.1. Dependent Variable

The idea of the dependent variable used in this chapter originates initially, as indicated earlier, from the Community Innovation Survey of 1997. This variable differentiates firms according to their internal capability to develop innovations. The exact wording of the question is as follows: "... who developed these products (or processes)? Mainly other enterprises or institutes? Your enterprise and other enterprises or institutes? Mainly your enterprise?" In this chapter, we use this question as an indicator of the creativity of innovative firms and this dependent variable takes a value of 1 when products and processes innovations are developed jointly by the firms interviewed and other firms or institutions (joint creativity); it takes a value of 2 when products and processes innovations are developed mainly by the firms interviewed (internal creativity); and finally it takes a value of 3 when no product or process innovations were developed by the firms interviewed during the last three years (no creativity). The alternative "innovations developed mainly by other firms or institutions" is not used in the present study because we did not have enough cases to use it for statistical analysis. The three alternatives of creativity are 1, 2 and 3 with 1 being the choice of a firm to develop or improve products or processes innovations jointly with other firms or organizations; 2 the choice to develop or improve products or processes with its internal resources; and 3 to be non creative firms, identified as the reference choice in our model.

The probability of choosing creativity category k (k=1,2,3) is given by:

$$\mathrm{Prob}_{ik} = \frac{e^{\beta_k X_i}}{1 + \sum_{k=1}^{3} e^{\beta_k X_i}} \qquad (1)$$

Where Xi is the matrix of creativity attributes (Appendix 1) and β_k is m x 1 vector of parameters.

As in the case of bivariate logit models, coefficients for reference choice are set equal to zero. Such normalization will be taken into account when interpreting the rest of model coefficients. In our case, the no creativity strategy is taken as a reference category and, as a consequence, estimated parameters will be interpreted in the following sense:

$$\frac{\text{Prob}_{i1}}{\text{Prob}_{i3}} = \frac{e^{\beta_1 X_i}}{e^{\beta_3 X_i}} = e^{(\beta_1 - \beta_3)X_i} = e^{\beta_1 X_i}$$

and (2)

$$\frac{\text{Prob}_{i2}}{\text{Prob}_{i3}} = \frac{e^{\beta_2 X_i}}{e^{\beta_3 X_i}} = e^{(\beta_2 - \beta_3)X_i} = e^{\beta_2 X_i}$$

or

$$\text{Ln}(\frac{\text{Prob}_{i1}}{\text{Prob}_{i3}}) = (\beta_1\text{-}\beta_3)X_i = \beta_1 X_i$$

and (3)

$$\text{Ln}(\frac{\text{Prob}_{i2}}{\text{Prob}_{i3}}) = (\beta_2\text{-}\beta_3)X_i = \beta_2 X_i$$

From (3), the estimated coefficients, for instance, β_{2j} (j = 1,….,m), are interpreted as the marginal change in the logarithm of the odds of the choice of the firm to develop or improve products or processes with its internal resources over the choice to be non creative due to a marginal change in the attribute j. However, while marginal changes in the logarithm of the odds are not always intuitively understandable, we can use the exponential of parameters which offers a straightforward model interpretation. In the previous example, if attribute j is a continuous variable, such as Percentage of Sales dedicated to R&D, $\exp(\beta_{2j})$ is the factor of change in the odds of the firm to develop or improve products or processes with its internal resources, instead being non creative, due to a one unit increase in the attribute j (Percentage of Sales dedicated to R&D). Hence, if $\exp(\beta_{2j})$ is higher than 1, it means that such odds will increase $\exp(\beta_{2j})$ times once Percentage of Sales dedicated to R&D increases in one unit. On the contrary, if $\exp(\beta_{2j})$ is less than 1, the same odds will decrease by such factor. In an analogous way, if attribute j is a dummy variable, (e.g. the high-technology sector where the variable takes the value 1 if firms operate in a high-technology sector, and 0 otherwise), then the estimated parameter β_{2j} suggests that the odds for firms operating in a high-technology sector to develop or improve products or processes with its internal resources, instead of to be non creative, is $\exp(\beta_{2j})$ times as high as the same odds for firms operating in a traditional sector. Hence, in the case of dummy variables, the exponential of parameters measures the factor of change in the odds with respect to the reference variable.

4.2. Independent Variables

To identify the explanatory variables of the creativity of manufacturing firms, we have estimated by maximum likelihood, for product innovations and process innovations, two sets of multinomial logit regressions using the different alternatives of creativity as dependent variables and as independent variables, the list of variables discussed in the second section of this chapter. Appendix 1 provides the overview of the operationalizations for the explanatory variables. The Cronbach's μ for all the independent variables based on multiple item scales is shown in Appendix 2. The values of the μ coefficients reported in Appendix 2 indicate that all but one of the multiple item scales employed in this study are reliable. The coefficient of the index of pressures from competition is too low to be considered as reliable.

5. RESULTS

Results of the regression of the creativity decisions of the firms, that is the decision about whether to develop or improve innovations jointly with other firms or organizations, to develop or improve innovations with its internal resources, or to be non creative, identified as the reference choice in our model, are summarized in Table 1 and Table 2, respectively for product innovations and process innovations. The two equations have good predictive power, with respectively 79.5 percent and 74.7 percent of correct predictions. The value of the Cox & Snell R^2 is .55 and .42 respectively, which is quite reasonable for qualitative dependent variable models. Furthermore, the computed values of the likelihood ratio (i.e., 236.6 and 163.8) are much larger than the critical value of the chi-squared statistic with 28 degrees of freedom at the 1 percent level. This suggests that the null hypothesis, that all the parameter coefficients (except the intercept) are all zeros, is strongly rejected. Consequently, the two models are significant at the 1 percent level.

The likelihood of joint creativity in products increases as the ratio of technicians and engineers increases, as firms increase the intensity of their relational assets, as their business network and information network increase, and as the degree of institutionalization of import of knowledge increases. The negative relation between the variety of advanced technologies and the likelihood of joint creativity means that as firms increase the number of different advanced technologies they use in their manufacturing processes, the likelihood of joint creativity decreases. Two other variables have shown a negative impact on the likelihood of joint creativity, the variable high-technology sector and the variable barriers to knowledge exchange. This negative relation means that firms operating in high-technology sectors are less likely to engage in joint creativity than firms outside the high-technology sectors. Finally, the estimation of this model also shows that as barriers to knowledge exchange increase, the likelihood of joint creativity of firms decreases. The meaning of these unexpected negative relations will be discussed in the next section.

The last two columns of table 1 show the coefficients and their exponentials for variables explaining the likelihood of internal creativity versus the likelihood of no creativity at all. We can see that the likelihood of internal creativity of firms increases as the percentage of sales dedicated to R&D increases, as the percentage of sales issued from exportations increases, as business network, information network and research network increase. Conversely, this likelihood decreases as the variety of advanced technologies increases, as the degree of pressures from competition originating from other firms increases. The likelihood of internal creativity also decreases when firms operate in high-technology sectors.

The results of the estimation for the factors influencing creativity in the matter of development or improvement of manufacturing processes of firms are shown in Table 2. The likelihood of joint creativity in the matter of manufacturing processes decreases with the increase of the ratio of sales dedicated to R&D and the ratio of sales shipped outside Canada. Conversely, the likelihood of joint creativity increases with the participation in business and information networks. As for the likelihood of internal creativity, it increases with the variety of advanced technologies, the participation in business networks, and research networks. However, the likelihood of internal creativity of firms in the matter of processes is negatively related to competitive pressures from other firms and to operation of firms in high-technology sectors.

6. DISCUSSION AND CONCLUSION

The results of the estimations for the factors influencing creativity in the matter of development or improvement of products and manufacturing processes point to the fact that different factors explain joint creativity and internal creativity as well as creativity in the matter of products and manufacturing processes. Moreover, some of these results call for discussion because they appear - at least at first sight - to be counterintuitive. We will discuss the results of this study by considering the different categories of variables by firstly considering the dependent variable.

The choice of the variable creativity as a dependent variable appears a fruitful choice given that the results of the two multinomial logit models revealed that different alternatives of creativity in the matter of products and processes were predicted by different explanatory variables. Furthermore, counterintuitive results regarding the negative relation between R&D, use of advanced technologies, operation in high-technology sectors and pressures from competitors can be considered as an invitation for more reflexion regarding the relation between creativity and its explanatory variables.

The creation of knowledge through codified business practices was measured with the percentage of sales dedicated to R&D. As expected, we have found that this form of creation of knowledge was positively related to internal creativity in the matter of product innovations. However we have also seen that this explanatory variable was not significantly related to the likelihood of joint creativity of firms in the matter of product innovations. As for the negative relation between R&D and joint creativity of firms regarding manufacturing processes, it suggests that joint creativity induces firms to assume that investments in R&D will be supported by their partners. Needless to say that if all firms expect that all their partners but them will invest in R&D, it will generate a situation where no firms will invest in R&D. A first contribution of this study is to point to the fact that the relation between R&D and innovation might be much more complex than suggested in models where innovation is measured with a question regarding the development or improvement of products or processes during the last three years.

The import of knowledge through codified business practices was measured with two indicators. The first measure took the number of different advanced technologies as an indicator of the variety of knowledge embodied in equipment. The results have shown, as expected, that there is a positive relation between this variable and creativity of firms in the matter of manufacturing processes. However, the results of the estimation for the impact of this variable on creativity of firms in the matter of products indicated that the variety of advanced technologies is negatively related to both joint and internal creativity. Although this result looks at first sight counterintuitive, we suggest that it makes perfect sense for two reasons: firms that have made significant investments in the acquisition of advanced technologies have higher incentives to maximize their investment by producing more of the same products than by developing or improving products that would require other significant investments in the acquisition of advanced technologies. The second reason, which is more speculative, is that firms that have a larger variety of advanced technologies might have higher incentives than others to rely on state of the art technologies and, as a consequence, might have lower incentives to develop or improve their products. This result suggests that the relation between advanced technologies and innovation might be more complex than suggested by studies taking innovation as their dependent variable.

The import of knowledge through codified business practices was also measured with a second indicator, the degree of institutionalization of the mechanisms used by firms to acquire knowledge usable for developing or improving products or processes. This second form of import of knowledge was found to have no significant impact on the creativity of firms in the matter of manufacturing processes. However, and as expected, the degree of institutionalization of import of knowledge was found to be positively related to joint creativity of product innovations, and not significatively related to internal creativity in the matter of product innovations. This result suggests that the institutionalization of import of knowledge is a lever of action that one should reserve only for the fostering of joint creativity in product innovations and that one should not use this lever to attempt to foster the other forms of creativity. Once more, this result suggests that the relation between the institutionalization of import of knowledge and innovation is far more complex than expected, thus complicating the life of policymakers whose mandate is to foster the creativity of firms.

The import of knowledge through uncodified business practices was measured with five indicators of social capital: participation assets, relational assets, business network, information network and research network. Overall, the results suggested that the import of knowledge through these diverse forms of social capital is less frequently related to creativity in the manufacturing processes than in product innovations, and that different forms of social capital explain creativity in product innovations. Hence, it makes sense to find that relational assets are significantly related to joint creativity in the matter of product innovations, but are not significantly related to the internal creativity of firms regarding product innovations. Likewise, one could expect that research networks are positively and significantly related to the internal creativity of products, but not significantly related to joint creativity in the matter of products. This result contributes to the advancement of knowledge by pointing to the fact that social capital does not affect equally all forms of creativity. Hence, social capital is a more efficient lever to foster the creativity of products than processes and, moreover, different forms of social capital are to be used to foster joint or internal creativity in the matter of product innovations.

The next variable considered was the capacity of knowledge management of firms measured by the percentage of technicians and engineers in the labor force of firms. Contrary to expectations, the results of the estimations of the multinomial logit models indicated that this factor is positively and significantly related to creativity only for joint creativity of product innovations. This counterintuitive result suggests that the relation between highly qualified manpower and innovation is more complex than usually assumed.

The barriers to knowledge exchange were measured with indicators related to the lack of cooperation between firms and diverse providers of information. This factor was found to have no significant relation for most forms of creativity, with the exception of one where we found that barriers to exchange of knowledge were negatively and significantly related to joint creativity of product innovations. This result makes perfect sense because it actually means that the decrease of these barriers would increase the likelihood of joint creativity of product innovations. Once more, this result suggests that the relation between barriers to knowledge exchange and innovation exert far more differentiated impacts than usually assumed in the scientific and policy literature on innovation.

Finally, and contrary to expectations, the results indicate that firms in the high-technology sectors are less likely to be creative than firms outside the high-technology sectors. This counterintuitive result is likely due to the fact that in a region where most firms are small, many firms classified in the high-tech sectors are actually more active in assembling products than in developing or improving products or manufacturing processes. The implication that we derive from this result is that the variable high-tech sector might not measure what we think it really measures. In other words, the concept of high-tech sector might not be a valid measure because there is not necessarily a perfect correlation between the idea of high-tech sector and the idea of high-tech firms.

This chapter addressed one question: Do business practices in the matter of knowledge management determine the creativity of manufacturing firms? The answer is yes, but the contribution of this chapter resides in the fact that it shows that diverse forms of creativity are explained by diverse forms of codified and uncodified business practices in the matter of knowledge management. Overall, the use of the creativity of firms instead of innovation generated much more nuanced results than the studies based on the variable innovation. More specifically, the results of this study suggest that factors that are usually associated to innovation have differentiated impacts on diverse forms of creativity, thus suggesting that more theoretical and empirical work needs to be done to improve our knowledge regarding innovation and the creativity of firms. Future research on innovation should attempt to improve the operationalization of the degree of novelty of innovations as well as to improve the operationalization of variables related to knowledge and knowledge management. Such a research agenda would be likely to improve our understanding of innovation and our empirical base to develop differentiated innovation policies.

Table 1.Multinomial Logit estimation for the factors influencing creativity in the matter of development or improvement of products of manufacturing firms

Dependent Variable: Creativity of the Firm

Independent Variables:

	Joint creativity/No creativity		Internal creativity/No creativity	
	Coefficient (β)	Exp (β)	Coefficient (β)	Exp (β)
Intercept	-1.358		-.994	
Percentage of Sales dedicated to R&D (PERR&D)	-.0034	.997	.0297 (.096)*	1.030
Variety of Advanced Technologies used (NBTADV)	-.207 (.009)***	.813	-.119 (.021)**	.888
Ratio of Technicians and Engineers (RATINTEC)	2,507 (.009)***	12.270	.242	1.274
Participation Assets (PARASS)	.047	1.048	.031	1.032
Relational Assets (RELASS)	-.293 (.006)***	.746	-.037	.964
Business Network (BUISNET)	.302 (.000)***	1.353	.172 (.001)***	1.188
Information Network (INFONET)	.149 (.07)*	1.160	.102 (.089)*	1.107
Research Network (RESNET)	.119	1.127	.126 (.066)*	1.135
Financial Assets (SALES)	0.00	1.000	0.00	1
Marketing Assets (EXPORT)	.015	1.015	.018 (.007)***	1.018
Competitive Pressures (COMPET)	-.140	.869	-.118 (.053)*	.869
Degree of Institutionalization of Import of Knowledge (INSTITU)	.480 (.035)**	1.616	.141	1.152
High-Technology Sectors (HIGHTEC)	-1.286 (.063)*	.276	-1.016 (.038)**	.362
Barriers to Knowledge Exchange (OBSTACLE)	-.138(.09)*	.871	.058	1.059

Number of cases : (Total = 297) 26 / 130 141 / 130
Cox & Snell R^2: .549Chi-square (d.f.): 236.6 (28) Overall percentage of correct predictions: 79.5%
a Figures between parentheses indicate p-value.
b *; ** and *** indicate that variable is significant at 10%, 5% and 1%, respectively.
c See appendix 1.

Table 2. Multinomial Logit estimation for the factors influencing creativity in the matter of development or improvement of manufacturing processes of firms

Dependent Variable: Creativity of the Firm

Independent Variables:

	Joint Creativity/No Creativity, Coefficient (β)	Exp (β)	Internal Creativity/No Creativity Coefficient (β)	Exp (β)
Intercept	-4.63		.087	
Percentage of Sales dedicated to R&D (PERR&D)	-.092 (.051)*	.912	.008	1.008
Variety of Advanced Technologies Used (NBTADV)	.083	1.087	.170 (.000)***	1.186
Ratio of Technicians and Engineers (RATINTEC)	.040	1.041	-.623	.536
Participation Assets (PARASS)	.030	1.031	-.016	.984
Relational Assets (RELASS)	-.104	.901	-.058	.943
Business Network (BUISNET)	.154 (.021)**	1.166	.196 (.000)***	1.217
Information Network (INFONET)	.171 (.030)**	1.187	-.0027	.997
Research Network (RESNET)	-.025	.975	.106 (.06)*	1.112
Financial Assets (SALES)	0.00	1.00	0.00	1.00
Marketing Assets (EXPORT)	-.026 (.018)**	.975	-.007	.993
Competitive Pressures (COMPET)	.085	1.088	-.171 (.000)***	.843
Degree of Institutionalization of Import of Knowledge (INSTITU)	.247	1.280	-.043	.958
High-Technology Sectors (HIGHTEC)	-1.00	.368	-.946 (.029)**	.388
Barriers to Knowledge Exchange (OBSTACLE)	-.028	.972	-.037	.963

Number of cases: (Total = 296) 19 /169 108 / 169
Cox & Snell R^2: .425Chi-square (d.f.): 163.8 (28) Overall percentage of correct predictions: 74.7%
a Figures between parentheses indicate p-value.
b *; ** and *** indicate that variable is significant at 10%, 5% and 1%, respectively.
c See appendix 1.

APPENDIX 1

Definitions of Independent Variables

Network Importance of the following information sources for the improvement and the development of new products and new processes (1= Not important at all to 5 = Extremely important)

Business Network Index:
- Clients;
- Suppliers;
- Competitors;
- Suppliers of equipments, material, components or software;
- Consulting firms; and
- Other firms belonging to your group.

Information Network Index:
- Fairs/Exhibitions;
- Professional congresses, meetings, and specialized publications;
- Internet and computerized data banks;
- Governmental information programs; and
- Patents documentation.

Research Network Index:
- Public research organizations (CRIQ, CNRC);
- Technology transfer organizations;
- Universities; and
- Community Colleges.

Relational Assets Index Level of consent with the following statements (1= Do not agree at all to 5 = Agree completely)
- I know personally professionals or managers in agencies of regional economic development;
- I know personally professionals or managers in government agencies involved in economic development;
- I know personally university or government researchers linked to the field of my products; and
- I know personally my clients and suppliers.

Participation Assets Index Participation frequency to meetings, associations or manufacturing industries networks according to the following levels (1 = Never to 5 = Very often)

- At the regional level;
- At the provincial level;
- At the national level;
- At the international level.

Degree of Institutionalization of Import of Knowledge Summation of the institutionalized means used to acquire information and technologies required to develop or improve products and manufacturing processes:

(1) Non exclusive licenses;
(2) R&D contracts;
(3) Acquisitions of equity in other firms;
(4) Codeveloppement of products and processes;
(5) Exclusive licenses;
(6) Joint ventures; and
(7) Acquisitions of other firms or mergers.

Pressures from Competition Index Level of consent with the following statements (1 = Do not agree at all to 5 = Agree completely)

- My clients can easily find a substitute of my products from my competitors;
- The arrival of new competitors is a constant threat;
- It is hard to keep employees and qualified workers;
- Our products become rapidly out of date;
- The production technologies evolve rapidly.

Financial Asset - The amount of the establishment's turnover during the last exercise.

Variety of Advanced Technologies Used - The sum of the number of technologies used by manufacturing industries within the following 21 advanced technologies:

(1) Computer Aided Design/Engineering (CAD/CAE);
(2) Computer Aided Design/Manufacturing (CAD/CAM);
(3) Modeling or simulation technologies;
(4) Electronic exchange of CAD files;
(5) Flexible manufacturing cells or systems (FMC/FMS);
(6) Programmable logic Contro (PLC) machines or processes;
(7) Lasers used in materials processing;
(8) Robots with sensing capabilities;
(9) Robots without sensing capabilities;
(10) Rapid prototyping systems;
(11) High speed machining;

(12) Near net shape technologies;
(13) Parts identification for manufacturing automation;
(14) Automated storage and retrieval systems;
(15) Automated vision-based systems used for inspection/testing of inputs and/or final products;
(16) Other automated sensor-based systems used for inspection/testing of inputs and/or final products;
(17) Local area network for engineering and/or production;
(18) Company-wide computer networks (including Intranet and WAN);
(19) Inter company computer networks (including Extranet and EDI;
(20) Manufacturing Resource Planning (PRPII) Enterprise Resource Planning (ERP); and
(21) Computers used for control of the factory floor.

Barriers to Knowledge Exchange Importance of the following obstacles to access information regarding the development or improvement of products and processes (1 = Not important at all to 5 = Extremely important)

- Lack of cooperation with other firms;
- Lack of cooperation with government research laboratories;
- Lack of cooperation with universities and community colleges; and
- Lack of information about technologies pertinent for developing or improving products and processes.

Percentage of Sales Dedicated to R&D - Percentage of the sales invested in R&D.

Marketing Assets - Percentage of the sales shipped to the United States and the rest of the world.

Ratio of Technicians and Engineers - Number of technicians and engineers divided by the total number of establishment's employees.

High-technology sectorsBinary variable coded 1 if the firm operates in a high-technology sector and 0 otherwise: The high-technology sectors considered are:

- Biotechnologies;
- Information Technologies;
- Aerospace;
- Instrumentation; and
- New Materials.

APPENDIX 2

Internal reliability coefficients (Cronbach's alpha) for variables including multiple item scales

Names of Variables	Number of cases	Number of items in scales	a
Relational Assets (RELASS)	440	4	.74
Participation Assets (PARASS)	440	4	.92
Business Network (BUISNET)	440	6	.90
Information Network (INFONET)	440	5	.91
Research Network (RESNET)	440	4	.92
Competition (COMPET)	440	5	.35
Degree of Institutionalization of Import of Knowledge (INSTITU)	440	7	.73
Barriers to Knowledge Exchange (OBSTACLE)	440	4	.79

Chapter Five
KNOWLEDGE FLOWS FROM PUBLIC INSTITUTIONS TO FIRMS

Michael Bordt
SIEID, Statistics Canada

1. INTRODUCTION

From existing information, what can we say about the flows of knowledge from the public sector to firms? How well is it managed and what are the impacts? Granted, the data sources are incomplete and often secondary but they do provide some insights into the importance and practice of KM.

A complete picture of flows of knowledge from the public sector to the private sector would require an understanding of the generation, codification, transfer, use and impacts of knowledge. This is not necessarily a linear model since using knowledge often leads to new knowledge. The following review is not intended as a summary of the current understanding of KM. It is an overview of concepts and assumptions that will bring some cohesion to the following discussions of intellectual property and data relating to KM issues.

Knowledge can be generated by research but, more and more, it is being realized that innovative ideas can come from many different sources including a firm's employees. How and where knowledge is generated is of interest. The increasing importance of collaborative research (through formal and informal consortia, as well as contracting) makes it a challenge to determine the precise source of new ideas.

Knowledge is codified in research papers, books, other documents and databases. Applying for a patent is one specific example. In this case, the inventor or innovator may not be the person who recognizes the potential benefits of the invention or innovation. Universities, government research labs and firms sometimes monitor the activities of researchers with the intent of identifying inventions and innovations with commercial potential.

Codified knowledge can be transferred and there are many mechanisms: reading, teaching, hiring (and losing) individuals with knowledge, licensing, formal and informal discussions are some examples. Technology often enters into the discussion at this point. With high bandwidth communications, the Internet, personal computers and "groupware" it is becoming easier and faster to find and share information - and sharing information is an important part of transferring codified knowledge.

It may be said that tacit knowledge cannot be transferred, it must be experienced. Mechanisms involved in tacit knowledge transfer include apprenticeship, tutoring, personal networks and on-the-job training. Some firms find it useful to maintain personal contact between distant team members (after first establishing trust with face-to-face meetings) using video conferencing and other high bandwidth technologies.

Codified knowledge can, to some degree, be managed. Efforts can be made to identify, codify, store and transfer it. Universities technology transfer offices are assigned the task of managing patents and licenses. Firms may have KM officers that encourage knowledge sharing between work groups or project teams. KM practices may be more or less formal. If KM is applied formally, a firm may have a knowledge strategy and full-time knowledge officer. However a firm may have been practicing good knowledge management without attaching the KM label to its practices. Hiring someone for who she or he knows, hiring for team building capabilities, having corporate apprenticeship, training and recruitment programs are some examples of this.

Flows of knowledge from public institutions to firms can take many forms. A researcher producing a commercializable invention, which is then patented and licensed to a company is an example of the traditional linear understanding of this flow. In fact, the researcher could be working under contract or in collaboration with the firm. The invention may be patented by the firm or by the university, depending on the contractual arrangements. The researcher may also decide to publish a paper on the topic without patenting and the firm could incorporate the invention into a product without licensing its use.

Another flow of knowledge from public institutions to firms is through the hiring of people. As students pass through the education system, they acquire knowledge and experience that are then used in their places of employment. Again, this is not necessarily a linear process: employees can return to school to upgrade their skills or to learn new ones.

2. CURRENT DATA SOURCES

This section describes some of the data sources that are being used or could be used to address the broad set of activities and linkages that are useful to track the flows of knowledge.

When addressing flows, it is best to understand both outflows and inflows. For example, we can ask universities about the value and number of licenses with firms, but to understand the impacts of these licenses, we would need to ask the firms about how these licenses are used. For this reason, the summary table (Table 1) shows where we have information on the activities in the public and private institutions but also information on the flows between the two.

2.1. R&D Surveys

Statistics Canada conducts annual surveys of R&D activities in Canadian industry, private non-profit organizations, and the federal and provincial governments. These surveys focus on expenditures, personnel, sources of funds and socio-economic objectives of the R&D (Statistics Canada, 2001a).

The higher education sector is not surveyed for R&D activities but the activities are estimated from administrative data and other sources (Statistics Canada, 2001c).

2.2. IP Management Surveys

IP management surveys are conducted for the higher education sector (Statistics Canada, 2000a) and for federal government science-based departments. The surveys cover IP policies, expenditures on IP management, numbers of intellectual properties reported, the number protected, patents received, contract research, spin-off companies and licensing.

Table 1: Current Measurement activities relating to knowledge flows

Component	Topic	Public Sector	Private Sector	Flows
Generation	Reasearch and Development	R & D surveys of federal and prov. governments, estimates of higher education sector	RDCI, RDCPNP	R & D surveys of federal and provincial governments, estimates of higher education sector, RDCI, RDCPNP
	Invention disclosures	IPHE, IPFG		not applicable
	Contracting (in and	IPHE, IPFG	RDCI, RDCPNP	RDCI, RDCPNP
Codification	Patent applications	IPHE, IPFG		Not applicable
	Publications	Bibliometrics	Bibliometrics	
Transfer	Patents	IPHE, IPFG	SIM99	
	Licensing	IPHE, IPFG		
	Copyrights	IPHE, IPFG		
	Spin-off companies	IPHE, IPFG		
	Sources of ideas		SIM99	SIM99
	Collaboration	Bibliometrics, SIM99	Bibliometrics, SIM99	Bibliometrics, SIM99
	Education	Various education Statistics on enrolment Performance, Graduation		Census of Population, NGS
	International mobility	Various sources	Various sources	Various sources
	Employee Involvment	WES		Not applicable
Management	IP management	IPHE, IPFG		Not applicable
	Use of technologies	ICT, WES	ICT, WES	Not applicable

Notes:
Bibliometrics: The Observatoire des sciences et de la technologie (see text).
ICT: Information and Communications Technologies and Electronic Commerce Survey (1999)
IPFG: Intellectual Property Management annex to the Federal Science Expenditures and Personnel survey.
IPHE: Survey of Intellectual Property Commercialization in the Higher Education Sector (1998, 1999)
NGS: National Graduate Survey
RDCI: Research and Development in Canadian Industry
RDCPNP: Research and Development in Canadian Private Non-Profit Organizations
SIM99: Survey of Innovation in Manufacturing 1999
WES: Workplace and Employee Survey

2.3. The Survey of Innovation in Manufacturing, 1999

This is a recent example of several innovation-related surveys that have been conducted over the past decade. Others include:

- The Canadian Survey of Innovation and Advanced Technology, 1993;
- The Survey of Innovation, 1996 (Services);
- Innovation, Advanced Technologies and Practices in the Construction and Related Industries Survey (1999) (Statistics Canada, 2001d); and
- Survey of Advanced Technology in Canadian Manufacturing, 1998 (Statistics Canada, 1999c).

The surveys of innovation ask if the firm has introduced a new or significantly improved product or process within the past three years and whether that product or process is new to the firm, to the country or to the world. This allows the analysis of innovative firms as well as the novelty of the innovation. The surveys also ask the sources of new ideas, the importance of information from various sources and collaborators in innovation. Choices for sources of ideas, information and collaboration include universities, the federal government and provincial governments.

The Survey of Advanced Technology in Canadian Manufacturing 1998, asks firms about their use of advanced technologies (including "knowledge-based software") and business practices (including "cross-functional design teams"). It also asks about employee training

2.4. Bibliometrics

The Observatoire des sciences et de la technologie, with the assistance of Statistics Canada, has developed the capacity to derive indicators of scientific output, collaboration and degree of specialization from the Science Citations Index (Statistics Canada, 1998a). It is possible for example, to determine the proportion of scientific papers that were produce in collaboration between universities, governments and firms.

2.5. Information and Communications Technologies

The Information and Communications Technologies and Electronic Commerce survey (Statistics Canada, 2000d) was first conducted in 2000 on a sample of public and private sector establishments. It asks not only about the presence and absence of ICTs (personal computers, networks, Internet, Intranets, and Extranets) but also how the Internet is being used. The uses include (among others):

- To share or perform collaborative research and development;
- E-mail (electronic mail)* (including text messages excluding attachments);
- Information searches;
- Exchanging electronic files (include attaching documents using E-mail software); and
- Education / training (interactive access to course materials online).

2.6. Workplace and Employee Survey (WES)

The Workplace and Employee Survey (WES), is a new Statistics Canada – Human Resource Development Canada undertaking designed to provide an integrated view of the activities of employers and their employees (Statistics Canada, 2000b). The WES consists of both employer and employee components. Employers are sampled by physical locations. Employees are then sampled from employer-provided lists within each location. The survey enables researchers to link employer policies, practices and outcomes with employee characteristics, activities and outcomes.

Of particular interest are the employee involvement practices (EIPs):

- Employee suggestion program. Includes employee survey feedback;
- Flexible job design. Includes job rotation, job enrichment/redesign (broadened job definitions), job enrichment (increased skills, variety or autonomy of work);
- Information sharing with employees. For example, with respect to firm's performance, colleagues' wages, technological or organizational changes, etc. This implies that employees have some feedback on policies;
- Problem-solving teams. Responsibilities of teams are limited to specific areas, such as quality or work flow (i.e. narrower range of responsibilities than self-directed work groups);
- Joint labor-management committees. Include non-legislated joint labor-management committees and task teams that generally cover a broad range of issues, yet tend to be consultative in nature; and
- Self-directed work groups. Semi-autonomous work groups or mini-enterprise work groups that have a high level of responsibility for a wide range of decisions/issues.

2.7. The National Graduate Survey (NGS) and Other Education Statistics

The National Graduates surveys are conducted by Statistics Canada in partnership with Human Resources Development Canada. They are specifically designed to obtain information on the relationship between education/training and labor market activities, the long-term labor market experiences of graduates, the employment, earnings, occupation and additional educational experiences and qualifications of graduates (Statistics Canada, 1998b). Every four years, a cohort of new graduates is surveyed. Each cohort comprises on average 40,000 graduates.

The NGS is also used to track graduates who move to the United States (Statistics Canada and Human Resources Development Canada, 1999), thus providing some insight into the flows of knowledge out of the country.

2.8. Census of Population and International Mobility Statistics

The Census of Population includes questions on each individual's highest level of education, field of study and current occupation. These data have been used to develop an initial understanding of where people with certain education backgrounds are working (Statistics Canada, 1999a).

After most of the recent Censuses of Population, Statistics Canada has performed a follow-up of a sample of persons who had answered the previous Census but did not appear in the current one (Statistics Canada, 1999b). This is one approach to determine who has left the country. The follow-up to the Census of Population 2001 will contain specific questions concerning the level of education and occupation of persons leaving the country.

3. SELECTED RESULTS

3.1. Public Institutions and IP

3.1.1. Identifying IP

For an organization to account for it's IP (inventions, reports, trademarks, copyrights, educational materials, etc.) the creation of that IP needs to be reported by the employee to the organization. In Canada, the requirement to report on IP creation at the university is a matter of university policy. Only 27 of Canada's 84 universities have policies that require reporting of all inventions (Statistics Canada, 2000a). At the other extreme, 17 universities have policies that specifically do not require the reporting of inventions and another 11 have no policies on reporting of inventions.

For those universities that require or recommend reporting, most depend on the researcher coming forth to make an invention disclosure. While 10 percent of the universities actively solicit researchers for "opportunities for commercialization", 70 percent say that the researcher is "primarily responsible for recognizing the discovery and its potential."

For the federal government, there is legislation (The Public Servants Inventions Act, R. S. 1985 c. P. 32) that ensures that inventions created by public servants are reported to the Minister of the department.

Figure1. Location of Innovation Partners: Public Organizations

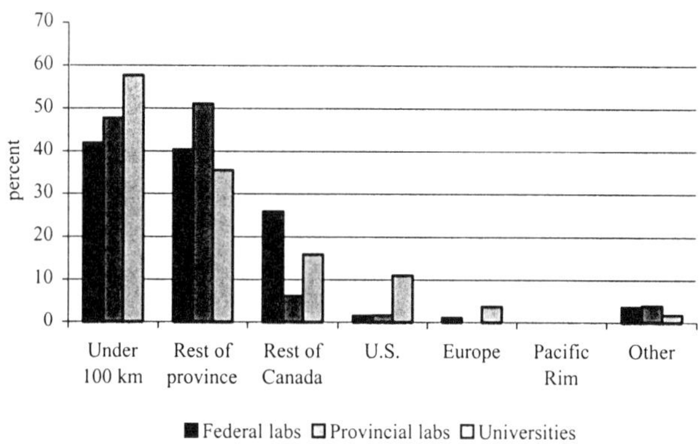

Figure 2. Location of Innovation Partners: Private Organizations

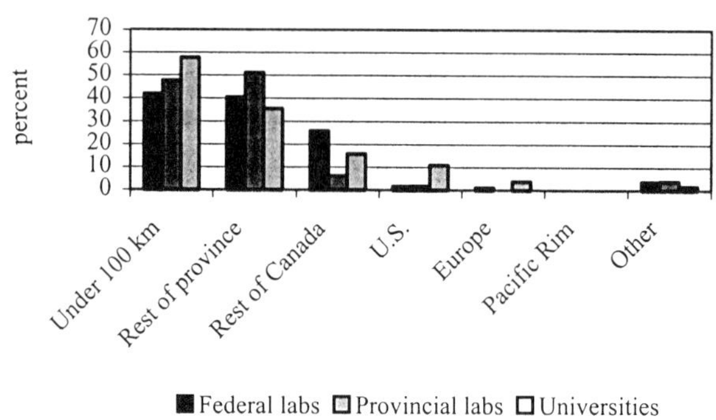

3.1.2. Educational Materials

With the advent of the Internet, professors have found it useful to post their course notes for use by students and colleagues. An emerging issue (OECD, 2000c) is that these notes are sometimes being adopted by commercial organizations without giving credit to the source. Technically, these materials are copyrighted by default but only about 10 percent of Canadian universities have policies to require reporting (and possible copyright registration) of these materials.

3.2. Collaboration and Innovation

3.2.1. Distance of Collaborators

Statistics Canada's Survey of Innovation 1999 (Statistics Canada, 2001) asks firms if they have brought a new product to market or have implemented a significant new process within the past three years. Those who have are termed "innovative firms". One third of innovative manufacturing firms in Canada report that they have developed these new products and processes in collaboration with partners. The reasons for this collaboration are to have access to:

- Critical expertise;
- Research and development capacity; and
- New markets.

When looking at this in more detail, it is evident that distance is important for certain types of collaboration. Firms tend to collaborate with nearby public institutions (universities and government laboratories, see Figure 1) but they collaborate with other firms irrespective of distance (Figure 2). This implies that, for sharing expertise, it is important to be near the collaborator.

3.2.2. Types of Collaborators

When viewed as a whole, according to the results of the Survey of Innovation, 1999, about 9 percent of innovative firms responded that universities constituted an important source of ideas. For those who have reported world first innovations, however, over 15 percent report that universities are an important source if ideas. World first innovative firms are also more than twice as likely to find federal and provincial laboratories as important sources (Figure 3). This also emphasizes the importance of universities to sharing expertise in support of the innovation process.

Figure 3. Sources of Ideas for Innovation

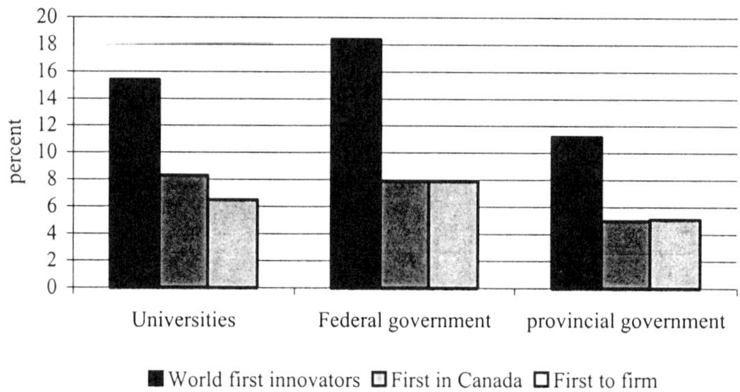

3.3. Employee Involvement Practices

According to Statistics Canada's Workplace and Employee Survey (Statistics Canada, 2001b), 65 percent of Canadian workplaces engage in at least one employee involvement practice. These practices include: employee suggestion programs, flexible job design, information sharing with employees, problem-solving teams, joint labor-management committees, and self-directed work groups.

3.4. Mapping Knowledge

Statistics Canada collects data based on it's hierarchical Standard Geographic Classification that divides the country into provinces, Census Divisions (CDs), Census Metropolitan Areas (CMAs), Census Sub-divisions (CSDs) and Enumeration Areas (EAs). Data from the Census of Population or larger surveys will be reliable at the EA or CSD level. Data from smaller sample surveys will only be reliable at the national or provincial level. Smaller aggregations would be based on too few responses to be reliable or non-confidential.

A geographic information system (or GIS) can be a powerful tool for integrating data that are based on different spatial frameworks. For example, single points (a university or a company) can be viewed in relation to a statistical area (for example, the CSD that it is in).

Statistics Canada has developed a prototype mapping application that includes public and non-confidential data on innovation and connectedness (Statistics Canada, 2000c). Many of the items used are derived from Statistics Canada's surveys on innovation, Research and Development in Canadian

Industry, and ICT and E-commerce. Several of the items derived from these sources are of interest to understanding knowledge flows, for example, the importance of government information and lack of university expertise as an obstacle to innovation (Figures 4 and 5).

The general patterns shown in the maps indicate that (a) government information tends to be more important in Quebec and outside of the urbanized and densely populated areas and (b) there appear to be several regions with lack of access to university expertise. More detailed analysis would be required to identify the reasons for these patterns.

4. DATA GAPS

There are two main areas of weakness in our current data: (1) knowledge and IP management activities of firms and (2) the links between public and private institutions.

Statistics Canada has not as yet undertaken a survey of IP management in firms that parallels the surveys conducted for universities and federal government departments. Such a survey would provide important new information on inventions patents and licenses. This would also provide better insight into the linkages.

There is also no official statistical information covering all the practices involved in knowledge management in either public or private organizations. Such a survey could provide details on which practices are employed (such as policies and strategies, incentives, hiring practices, training, and technologies).

Besides tracking the flow of IP from the public sector to firms, there is also a need to understand the flows of knowledge embodied in people. Better information on the skills, level of education and work experience of people as they move from universities to firms and between firms would be important to understand the links between public and private institutions.

5. CURRENT INITIATIVES

Statistics Canada is planning a Survey of Knowledge Sharing and Knowledge Transfer Practices during this calendar year. This will cover the key knowledge management practices as well as the sources of ideas and reasons for implementing these practices. The survey will likely be conducted in the fall of 2001 and results should be available in early 2002.

Figure 4. Firms Identifying Federal Government as an Important Source of Information
Source: SIEID, Statistics Canada

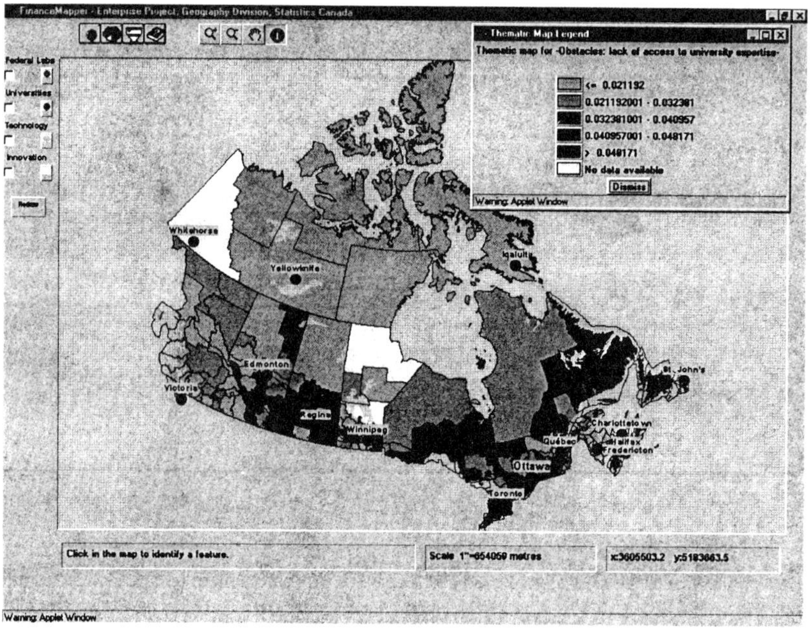

Figure 5. Firms Lack of Access to University Expertise as an Obstacle to Innovation
Source: SIEID, Statistics Canada

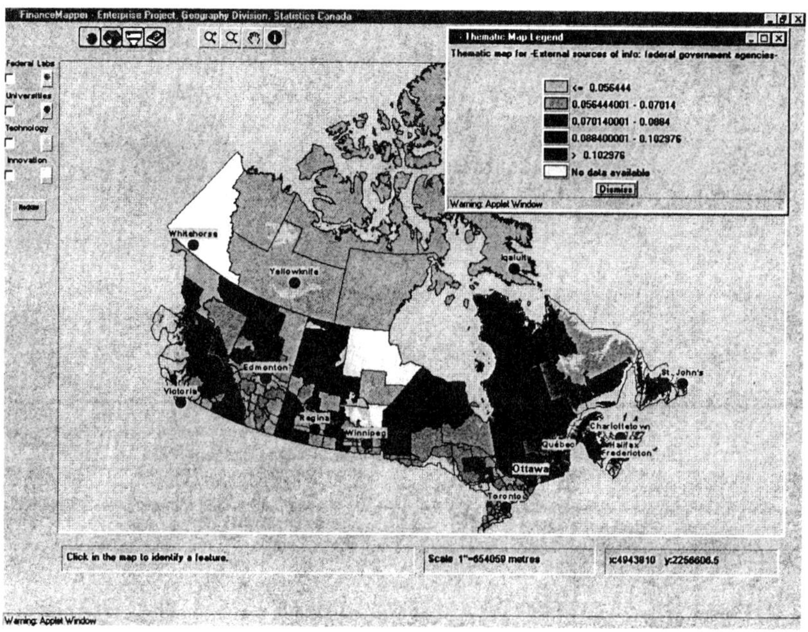

APPENDIX 1. DEFINITIONS AND CONCEPTS

Intellectual property covers the variety of "creations of the human mind" that can be protected by law:

Inventions: Any patentable product, process, machine, manufacture or composition of matter, or any new and useful improvement of any of these, such as new uses of known compounds (Canadian University Intellectual Property Group, 1998). Some inventions are patentable in some jurisdictions but not in others: these include novel genetically-engineered life forms, new microbial life forms, methods of medical treatment and computer software.

Computer Software or Databases: As noted above, computer software can be patented but normally it is protected by copyright. Databases may also be copyrighted.

Literary, Artistic, Dramatic or Musical Works, Books, Papers: This category includes any copyrightable works other than computer software and databases and special educational materials as noted below.

Educational Materials: This category includes special materials that may be copyrighted but are not necessarily in the form of printed books. This could include broadcast lessons, Internet pages, booklets, posters or computer files, among others.

Industrial Designs: These are original shapes, patterns or ornamentations applied to a manufactured article. Industrial designs are protected by registration with the Canadian Intellectual Property Office.

Trademarks: These are words, symbols, designs, or combinations thereof used to distinguish your wares or services from someone else's. Trademarks are registered with the Canadian Intellectual Property Office.

Integrated Circuit Topographies: This is a three-dimensional configuration of the electronic circuits used in microchips and semiconductor chips. Integrated circuit topographies can be protected by registration with the Canadian Intellectual Property Office.

New plant Varieties: Certain plant varieties that are new, different, uniform and stable may be protected by registration with the Canadian Intellectual Property Office.

Know-How is practical knowledge, technique or expertise. For example, certain information is codified in the patent application but a researcher's know-how could be valuable for commercial optimization of the product. Know-how can be licensed independently of the terms of a related patent.

Most of these "intellectual properties" can be protected using one or more of the existing legal frameworks:

Patents: A patentable invention, to be protected, requires a patent application with the government of the countries in which protection is sought. A patent application may be preceded by an invention disclosure to the institution.

Registration of Copyrights: The kinds of works covered include: books, maps, lyrics, musical scores, sculptures, paintings, photographs, films, tapes, computer programs and databases. A copyright means that the owner is the only person who may copy his or her work or permit someone else to do so. You obtain copyright automatically in Canada when you create an original work. It is not necessary to register copyrights with the federal Copyright Office but doing so can be a proof of ownership.

Industrial design registration gives the owner exclusive rights to use the design. The design must be an original shape, pattern or ornamentation applied to a manufactured article.

Trademark registration gives the owner exclusive rights to words, symbols and designs, or combinations of these, that distinguish one's wares or services from those of someone else. Trademarks are registered through Canada's Trade-Marks Office. Normally, trademarks do not need to be registered, however, doing so gives the owner exclusive rights throughout Canada.

Integrated circuit topography registration gives the owner exclusive rights to use the design. Protection can extend to the layout design as well as to the finished product.

Plant breeders' rights gives the holder exclusive rights to new varieties of some plant species. To be protected, the varieties must be new, different, uniform and stable. A claim for protection is preceded by publication of a description of the plant variety in the Plant Varieties Journal.

Non-disclosure or confidentiality agreements: Non-disclosure is an alternative to patenting. A non-disclosure agreement does not constitute a property right although the IP can be protected by contract. Parties to a non-disclosure contracts agree not to divulge valuable technical knowledge and can be prohibited from so.

Chapter Six
KNOWLEDGE MANAGEMENT IN SMALL FIRMS:
Theoretical Perspectives and Evidence

Hans G. Schuetze
University of British Columbia

1. INTRODUCTION

A recent headline in a major German newspaper read 'Many companies pay lip service only to knowledge management - have no knowledge about their own competencies and knowledge is not shared with staff'. The article[1] referred to a research study that surveyed top managers of 112 of Germany's largest 500 companies with respect to their companies' practice of knowledge management[2]. The study found a huge gap between the perceived need for knowledge management and the companies' actual practice. Thus for instance, 98 percent of the management believes that regular systematic environmental scans about future trends and developments are a strategic necessity, yet only 20 percent are conducting such comprehensive scans. Only 28 percent of the companies communicate to their own staff the results of their scans. Data from outside the company flows primarily to top management, strategic planning, and sales, but not to the research and development (R&D) and production units. Knowledge about the companies' own competencies is underdeveloped as only 26 percent of firms are looking into the effects of personnel fluctuation and thus the effects of the drain of the firm's knowledge base.

What picture would emerge if a study on knowledge management were to be conducted in small firms[3]? What would be different with respect

---

1. Frankfurter Allgemeine Zeitung, Oct 30, 2000

2. Future-oriented knowledge management in German companies? An empirical study by Witten-Herdecke University and the Market Lab AG (June 2000).

3. 'Small firms' are defined differently depending on the jurisdiction. In most systems the number of employees defines smallness. The OECD distinguishes very small firms (0 - 9 employees), small firms (10 - 99), medium-sized firms (100 - 499) and large firms (500 +). For the purpose of a study on knowledge management, one would probably have to eliminate the very small category, but include the lower portion of the medium-sized category. Thus Davenport and Prusak (1998) hold that "the maximum size of an organization in which people

to the flow, creation and use of knowledge in small, localized companies where "a manager probably knows who has the experience in a particular aspect of the business and can walk across the hall and talk to him" when they are compared to large, global "enterprises with scattered offices and plants and a complex mix of products and functions" (Davenport and Prusak, 1998, p. 17 and 18). It is the premise of this chapter that knowledge management is also important for small firms - even if they would probably not use the term or know what it means. However, for these firms knowledge management presents problems of another kind, in particular finding relevant information and know-how from outside the firm, and absorbing and applying it to the firm's business.

Much of the research on 'knowledge management' and on related concepts such as 'organizational learning' and 'learning organizations' has been conducted in larger industrial firms. Especially most of the literature written by management consultants concentrates on work organization, management, innovation, and learning processes in large enterprises, and most of them are research and development intensive or from the high-tech sector. The paucity of research on structures and processes in smaller firms can be explained by a number of reasons, especially the wide variety of business activities, the greater informality of management structures and styles, and probably also the fact that these smaller firms do not typically employ or seek the advice of management consultants. However such research would be of use as it would help managers, policy analysts, and institutions to better understand the specific problems that smaller firms typically face when operating under the conditions of a knowledge-based economy. This would be of importance since small firms play an important role in economic growth, job creation, and economic development and are therefore of great interest to policymakers (e.g. OECD, 2000d and 1993).

The few studies that focus on small and medium-sized enterprises (SMEs) show that many of them are as productive and innovative as their larger counterparts (OECD, 2000d; Acs and Audretch, 1990). These firms are competitive with larger firms because of their shorter communication channels within the firm, the more ad hoc nature of their decision-making, their flexibility in reacting to changing markets and new business opportunities, and their ability of exploiting information and knowledge from outside the firm. The 'innovative capability' is dependent of their ability of 'organizational learning', their 'absorptive capacity', their linkages to, and communication with external networks, and their capacity for managing information and knowledge from inside and outside the enterprise, but in particular from outside.

These concepts and capabilities are discussed below, as they seem particularly appropriate for explaining the mechanisms of knowledge

---

know one another well enough to have a reliable grasp of collective organizational knowledge is 200 to 300 people" (p. 17).

management in small firms. In the second part of this chapter, some empirical evidence from a study on small firms will be presented to demonstrate how these concepts apply to, and are exemplified by the practice of knowledge management in these firms.

2. THEORETICAL PERSPECTIVES

The term 'knowledge management' is relatively new. Its emergence as a management concept is the result of the recognition of 'knowledge' as an intangible yet very valuable corporate asset which needs systematic attention and careful managing in order to get the maximum value from it (Davenport & Prusak 2000). In a simple form, knowledge management means the process of collecting, organizing, classifying and disseminating information throughout an organization. In a wider sense it can be understood as the philosophy and technique of recognizing, increasing and exploiting the organization's intangible assets. This is the meaning it is given here.

Contributions to the discussion on knowledge management, organizational learning, the learning organization and other related concepts for SME's come from at least five disciplines. These different approaches reflect (1) the complex nature of knowledge management and the close connection and partial overlap with other theoretical constructs, and (2) the multi-disciplinary nature of these perspectives.

The first is organization theory which stresses the importance of the environment on organizational structures and development as well as the importance of structures. Sociology has contributed important insights into learning as a social process which is strongly dependent on networks, and into inter-active social processes in the workplace. Psychology has widened our understanding of various learning process, both individual and collective, and both formal and informal. Management theory draws on the former three disciplines when explaining 'how organizations manage what they know' (thus the subtitle of the book by Davenport and Prusak, 1998).

The fifth discipline is economic theory which has shifted from a neoclassical to an evolutionary perspective and from an exclusive focus on market behavior to one which includes the role of institutions in economic development. Economic theory explains why and how the on-going transformation from 'industrialist' to knowledge-based economy profoundly changes the ways in which firms do business (e.g. Lipsey, 2000; Edquist and Riddell, 2000). In the 'new economy', competitiveness is still dependent on a number of traditional elements, such as capital, market resources, technology, and the workforce. However, it is becoming increasingly based on less tangible factors, such as information, communication, knowledge, research, and the human resources of a company and their ability and willingness to communicate, share knowledge and learn. The new competitive advantage is based on the firms' capacity to innovate, to adapt quickly to changing demands, conditions and regulatory environments, and to take advantage of both existing and new market opportunities[4].

2.1. Innovation, Innovation Systems, And Networking

Innovation is not a single event, but an on-going process of "learning, searching and exploring, which result in new products, new techniques, new forms of organization and new markets" (Lundvall, 1995, p. 8). It is dependent on a mix of ideas, information, theoretical knowledge, and practical know-how. The speed of the innovative process differs not only because of different strategies and management styles that firms adopt, but also because of differences in the business environments, e.g. market demand, state-of-the-art technology, availability and price of raw materials, regulatory framework, business 'climate' etc. While some of these elements are identical or similar across the economy, others differ with respect to the sector to which the firm belongs, and the place where it is located.

It follows that the process by which innovation takes place is varied and complex. It does not follow a 'linear' path from basic research to the market. Rather, it is characterized by multiple linkages to the business environment, interdependent relations with others, and the various routes through which information and feedback flow. Firms do not operate in isolation, but communicate and cooperate, and often compete, with other firms; they have also linkages to various other organizations, public institutions, or mediating agents. Together these organizations and institutions form the firms' 'business environment', a 'system of innovation', in which they operate, and of which they are a part (Lundvall, 1995 and 2000; Edquist, 1997).

Because firms do no operate in isolation, but interact with others, linkages to the 'innovation system' and networks are important for companies of all sizes. But they are crucial for smaller firms since most of them do not have, unlike their larger counterparts, the personnel, expertise or facilities to access and use original sources of information and knowledge, such as specialized data banks or libraries. For small firms, networks form the conduits through which a great deal of information flows throughout the 'system'.

Much of the information businesses require to compete in the market, ranging from market data to technical information, comes through such networking. Suppliers, customers, competitors, public agencies, professional

---

4. The term 'innovation' is often used to describe technical novelty embodied in a product. Yet, following Schumpeter's (1934) classical definition, five different types of innovation can be distinguished: introduction of a new product or of a significant qualitative change in an existing product; process innovation which is new to the industry; the opening of new markets; development of new sources of raw materials or other inputs; and significant changes in workplace organization and management.

associations, and other agents constitute an enterprise's networks. In addition, there are personal networks to which the owners, managers and employees belong. These personal or informal networks are often equally important. Examples are memberships in professional associations, contacts with former colleagues in other firms or the public sector, former classmates, academic mentors, and the like. All of these contacts can be used by firms for systematic information gathering and scanning of the business environment - not just with respect to market conditions but also 'soft' data which can determine longer-term prospects and developments.

2.2. Information Flows And Organizational Structure

Changes in the business environment require enterprises not only to re-think their strategies with respects to products and markets, but also their work organization and practice. In order to react flexibly to changing market demand and customer specification and feedback, centralized management structures and a Taylorist division of labor does no longer work for many firms. Tasks that were distributed between different units are increasingly integrated in these firms and delegated to teams or groups who deal directly with customers' specifications and feedback. The organizational structure changes and organizations become 'flatter' as a result.

To be sure, the distinction between hierarchically organized and 'flat' organizations is not very precise. In principle, 'flat' connotes fewer levels of authority and control and a greater degree of decentralized decision-making whereas 'hierarchical' means several such levels and a tight control of information. Flat companies are characterized by intensified communication process and information sharing.

Whether a company adopts a hierarchical or a flat organizational structure is usually a reflection of the way information and knowledge are acquired and used, and in particular the flow of information within the firm. In order to permeate the entire organization, all information that might be relevant (widely defined) is distributed to everyone in the firm, moving beyond the immediate and present operational needs of employees to carry out their individual tasks. While much of this information may be considered 'redundant' or an 'overload', it enables all members of the staff to better understand the firm's 'core knowledge', i.e. its business strategy and directions, as well as the various environmental factors that do, or might, influence the firm. Such free flow of potentially relevant information is achieved through a variety of organizational arrangements, such as in-house meetings, seminars, and brainstorming sessions, or the sharing of information from academic or professional conferences and trade shows. These exchanges and this communication can be organized either on a regular or ad-hoc basis.

The type of organizational change firms choose depends partly on the organizational culture and the management style of the individual firm. It can also depend on the size of the firm. Small firms are not necessarily 'flat' or decentralized so that information can flow uncontrolled and unhindered. On the contrary, the strong position of the top manager who is also often the founder and the sole owner often imposes a hierarchical structure in very small firms. Even though there is very often in such firms an informal structure in place which facilitates information flows independent of the hierarchical structure, the flow of data and information from outside the organization tends to be controlled by the owner, or another designated person reporting to the owner, who act as 'gatekeepers' for outside information.

The decision to have intensive communication and uncontrolled information flows is also influenced by the firm's general business environment. If markets are relatively stable and the product mature, there tends to be less feedback or need for information, and organizations tend to favor a hierarchical form. This type of organization is typically found in manufacturing industries engaged in large batch production of standard products. By contrast, where markets are more volatile and production is small batch and customized, organizations tend to be more flexible and flat. In conjunction, where industries are high-tech or knowledge-based, relying on customer feedback and other external sources of information for their design and production process, employees are more likely to work in project groups or teams, rather than in vertically organized departments or divisions. Hybrid forms can also be found, for example matrix type organizations, where different elements are mixed, with vertical chains of command and horizontal cooperation between specialists from different departments and units.

Besides opening up communication channels and access to information, flat organizations mean also delegated decision-making power. Such delegation of authority and control increases employee autonomy, i.e. the possibilities for employees to act, either individually or in a group, as independently as circumstances permit. (Semi-) autonomous work teams are an example.

2.3. Organizational Learning

Learning is central to knowledge creation, absorption and use. Unlike formal training, learning is not a singular and isolated event, but the result of an ongoing communication process occurring as a part of everyday experience, and includes activities such as learning by doing, learning by using, and learning by interacting with others, both within and outside the firm.

Different types of knowledge require contrasting styles of learning, involving differing forms of communication and interaction. Organizational 'knowledge creation', or 'organizational learning', is a process by which an organization, through its internal mechanisms "amplifies" the knowledge created by individuals, and makes it part of the knowledge network of the organization.

Nonaka and Takeuchi (1995) distinguish four modes by which knowledge is created by passing it on to other members of a firm's workforce:

- Socialization is the process of sharing experiences trough observation, imitation and practice, thereby creating shared experience, mental models and technical skills. Examples of this are learning within teams ('communities of practice', Brown and Duguid, 1991), or certain forms of workplace training, such as apprenticeships and on-the-job mentoring;
- Externalization involves the articulation of personal or group experience and know-how. While occasionally this is done by an individual systematically recording and conceptualizing his or her experience, the conversion process typically involves either a dialogue with others or collective reflection;
- Combination or reconfiguration of different bodies of explicit knowledge occurs by putting together different known technologies or processes to new applications ('system integration'), thereby creating new knowledge or competencies ('know-how'); and
- Internalization is the acquisition of knowledge through documents, drawings, or oral stories thereby appropriating shared mental models or technical know-how.

To promote organizational learning five conditions are typically required: worker autonomy, free information flows, flat organizational structures, a learning strategy, and an appropriate stock of skills embodied in the workforce (Argyris and Schon, 1978). Worker autonomy and the flow of information in a flat organization were mentioned already.

While individual learning occurs almost automatically through the interaction with others in the workplace, for organizations to learn and create new knowledge, firms must have a strategy for "developing the organizational capability to acquire, create, accumulate and exploit knowledge" (Nonaka and Takeuchi, 1995). Such a strategy can be codified, e.g. in the form of mission statements, work organization plans, personnel development or human resource development strategies. Such a learning strategy comes often in the guise of a formal training policy or plan which sets the objectives for the provision of, and participation in, formal training activities. In the case of small businesses, learning strategies tend to be non-codified and implicitly contained in the overall business strategy. This means that the firm's strategy is not always understood, subscribed to in terms of a collective commitment, nor implemented.

Many smaller firms have no concrete learning strategy at all. Formal or informal training is organized ad hoc and when management sees a necessity for it. This is often the case when there is a new piece of equipment or technical system that employees need to be familiarized with in order to put it to an efficient use. Such event-triggered training and learning is, of course very different from the kind of systematic and sustained leaning activities that are the characteristic of 'learning organizations' (Betcherman et al., 2000).

Surveys on workplace training seem to support the impression that the incidence of training in small firms is considerable lower than in larger ones (Betcherman et al., 1998). While the statistical evidence cannot be disputed, I have argued elsewhere (Schuetze, 1998) that smaller firms tend to compensate for the lack of formal training by organizational arrangements that facilitate informal ways of training and learning which are not captured by traditional training surveys. While there is thus no direct statistical evidence for this thesis, there are some data from company surveys and considerable evidence from case studies (e.g. Schuetze et al., 1997) that support this thesis.

2.4. Human Resources Development In The Knowledge-Based Economy

There is ample evidence that the on-going change to a more knowledge-based economy is leading to increased demand for higher levels of technical and/or job-specific skills (e.g. Gingras et al., 2000). Companies are also looking for overall improvements in the more generic skill areas such as communication, problem solving, team work and adaptability.

This upward skill bias has a number of consequences for enterprises and their investment in human resource development. Faced with critical skill shortages, firms must decide what type of new skills to bring into the firm, either from the outside through hiring or making use of sub-contractors and consultants, or from inside by investing in the skill development of their existing personnel. In the latter case, firms can either offer formal training themselves, for which many SMEs do not have the resources nor the expertise, or send their workers to a training institution and pay for their tuition and incidental costs. Even if the required programs are offered by outside providers, which is not always the case, this involves costs that the firms are frequently unable to bear, especially if no money for training had been set-aside in the budget.

The evidence was already mentioned which shows that the incidence of formal training decreases with the size of firm and that therefore employees in smaller firms are less likely to receive the same amount of training as their counterparts in larger firms. But it was also pointed out that the lower incidence of 'training' does not necessarily imply 'less learning' in the workplace. Training in smaller firms cannot be neatly separated from the informal mentoring and learning activities that are integral part of the actual working process. In firms with a flat hierarchy, a clear-cut division of labor, and therefore clearly defined work tasks and skill requirements are often not possible. This blending of jobs often necessitates more informal, incidental learning than exists in clearly defined organizational structures typical of larger firms.

The work in teams that has been mentioned above is a form of work organization leading to more worker autonomy with regard to decision-making and control. Such autonomy is not only essential for motivating adult learners to learn, it is also conducive to the learning process itself (Rubenson and Schuetze, 1996). Learning is enhanced when these teams are not only able to organize their own work within the framework of the task assigned, but are also multi-functional, drawing members from a cross-section of the firms' activities. Such work teams work at different stages and levels, ranging from development and design, or production, to quality control, e.g. quality circles. Since the knowledge base of such groups is larger than that of the individual members, constant knowledge creation and learning take place through processes of social interaction with fellow workers. The various forms of this interaction and the corresponding modes of learning were already discussed above whereby the individual knowledge of various members is extended both on an individual and a collective level.

But the small size of the firms is also conducive to learning outside of such teams or working groups. Recurrent opportunities of various kinds to exchange information, experience and stories can be arranged without the formality of officially organized training sessions, and often occur as part of the regular workday. Examples range from reports to colleagues of information and contacts from conferences and trade shows to informal lunch breaks and talks in the coffee room. Of course, such informal incidents of information sharing are not occurring exclusively in small firms; however in large firms such unstructured and often spontaneous exchanges are not likely to reach all the recipients who might be able to use the information. To be sure, there might be inhibiting factors of another kind in small firms which prevent the free exchange of information and mutual learning events, such as differences in status, problems of specialized language, or lack of trust (Davenport and Prusak, 1998). However, given the proximity between the different offices or production sites and the possibility, in fact unavoidability of direct contacts between the staff members, these can probably be overcome more easily by a strategy that actively promotes regular information exchange and a learning culture throughout the company.

2.5. Absorptive Capacity

The ability of a firm to recognize, assimilate, and subsequently make use of new information from external sources has been called its 'absorptive capacity' (Cohen and Levinthal, 1990). It is critical to innovation in small firms since we have already discussed that they have to rely to a much greater extent than larger firms on such external sources.

External information is only retained when viewed as relevant and valuable, which ultimately depends on the perception, knowledge base and imagination of employees who have access to such information. The ability to recognize relevant information and knowledge is closely tied to the staff's prior knowledge and experience as well as their diversity of background. Yet, an organization's absorptive capacity not only depends on the sum of its members, it also reflects ability to exploit this external information and knowledge. This ability is typically related to the organizational structure of the firm, the flow of information within the firm, and the mechanisms by which information from external sources is acquired, converted into knowledge and used for commercial ends.

Since, initially, "all learning takes place inside individual heads" (Simon, 1991, p. 125), an important prerequisite of a company's absorptive capacity are the knowledge, skills and experience of the owner or manager and the employees. It seems obvious that small firms might be at a disadvantage here in comparison with larger enterprises which typically employ workers with more diversified backgrounds, thus enlarging the area of competency and potential absorptive capacity. However, this is not necessarily so. The knowledge critical for absorption of external information is "not simply … substantive, technical knowledge; it also includes awareness where useful complementary resides within and outside the organization. This sort of knowledge can be knowledge of who knows what, who can help with what problem, or who can exploit new information" (Cohen and Levinthal, 1990, p. 133).

Specialized small firms with a workforce of advanced skill levels will usually have the necessary knowledge base to tap into external sources of information that are potentially relevant for their company. But even those small firms that lack such prior related knowledge tend to have some critical knowledge of this latter kind, using business services, industrial or professional associations, or public support mechanisms, e.g. the Industrial Research Assistance Program (IRAP) of the National Research Council, which have been specially established to support the absorptive and innovative capabilities of small firms (OECD, 1995b). Also, they are usually not kept from seeking help from outside experts or support organizations by a 'not-invented here' syndrome that keeps many larger firms from accepting innovative ideas from outside.

2.6. Summary

Five concepts and themes were discussed that seem to be specifically relevant for interpreting the ways in which information and knowledge are managed in smaller firms.

(1) Starting with innovation systems theory, the linkages to, and networking with, the system or 'environment' were identified as crucial for

small firms as much of the information and knowledge needed cannot be found inside the firms themselves.

(2) Unhindered access to information within the firm was seen to be advanced by non-hierarchical structures that permit and require the free flow of all potentially relevant information. Flat organizations are also characterized by decentralization of decision-making and greater autonomy of workers. Semi-autonomous teams and project groups, and regular communication in cross-functional units, e.g. quality circles, are manifestations of such structures.

(3) Organizational learning and the ways in which such learning takes place were discussed with respect to different types of knowledge that reside in the firms' human resources. Communication and information sharing were identified as crucial elements of this process.

(4) The increased importance of human resources in a knowledge-based economy makes the knowledge and skills of the workers critical assets of the firm. As some of this know-how becomes outdated without further learning, the motivation and ability of workers to engage in further learning. The creation of a learning environment become central tasks for the firm that wishes to manage its knowledge base.

(5) Finally, the concept of 'absorptive capacity' establishes the connection between the information and knowledge residing outside the firm and the cognitive and organizational structures within the firm. This approach links the ability of a firm to recognize the value of new, external information, and to their ability to assimilate new information with prior related knowledge (embodied in the firm's workforce), and finally, to apply the combined knowledge towards commercial ends.

We shall now look into some empirical evidence from a study on innovation in small firms, trying to link these concepts and the evidence and to discuss the findings in the light of these concepts.

3. ELEMENTS OF KNOWLEDGE MANAGEMENT: EMPIRICAL EVIDENCE

To explore the complex nature of knowledge management and related concepts is beyond the scope of a survey which is directed to time pressed owners and managers of SMEs. However, data from a survey provide information about the nature of relevant structures and processes in small firms. It was pointed out before that knowledge management in smaller firms seems less complex than in large and globally operating enterprises where the stock of knowledge is scattered over several sites and organizational units. On the other hand, it is more difficult because a major part of the knowledge needed is not residing in the firm itself, but must be found, accessed, and appropriated from various outside sources and distributed and absorbed within the firm.

The survey was part of a larger study on innovation, organization, and learning in small firms in British Columbia (B.C.) (see Schuetze et al., 1997; Schuetze, 1998). It was preceded and followed up by a series of case studies. The study covered two manufacturing and two service-based sectors: secondary wood manufacturing, telecommunications equipment manufacturing, engineering consulting, and computer services. Firms included in the survey had between 10 - 100 employees and were in operation for three years or longer. Thus, very small and start-up firms were excluded on the one end, and larger firms with advanced levels of division of labor and professional management, on the other[5.].

In providing a brief analysis of the results, evidence from the case studies will be cited where this seems useful for the interpretation of the survey data[6.]. The analysis addresses four areas that are relevant to knowledge management: firm organization, linkages with the business environment and external sources of information, the innovation process, and human resources development and individual and organizational learning.

3.1. Organization

The 'organization' of a firm comprises the internal structures and modes of operation. As mentioned before, organization and the capacity to manage information and knowledge are closely related. One of the features that is typically attributed to organizations that are successful in managing information is that they are 'flat', as opposed to hierarchical in their structure.
Levels of Decision-Making (Hierarchy)

When asked how they would describe their organizational structure, 38 percent of firms classified their operations as hierarchical, and 57 percent as flat (Table 1). The two manufacturing sectors, telecommunications and secondary wood manufacturing, tend to have more hierarchical structures whereas the majority of computer and engineering consulting firms are seen by themselves as having flat structures.

---

5. The survey, carried out in 1997, was in the form of a mail back questionnaire with 38 questions. Companies were invited to provide additional information and to comment on the survey. Data on small firms of the appropriate size were provided by BCStats, several industry and professional associations, and the project advisory group which was composed of representatives of business and labour as well as several ministries and agencies. A total of 236 firms were randomly selected; the response rate was 50 percent. As some were incomplete or came from firms that were larger or smaller than the intended firm size, only 46 percent were used. In some selected cases, follow-up telephone interviews were used to solicit clarification or additional information.

6. For the full study, on which the following sections are based, see Schuetze et al. 1997.

Table 1: Work organization structure by size and sector (percent)

	Hierarchical	Flat	No Answer
Telecommunications	50	46	
Computers	24	76	
Secondary Wood	44	40	
Offices of Engineers	34	63	
Total*	38	57	6

* percentage does not add up to hundred due to rounding

Since the survey questionnaire provided no exact definition of the terms 'hierarchical' or 'flat' in terms of levels of decision-making and control, owners or managers responding to the questionnaire had to decide for themselves if their firms were 'flat' or hierarchical. A review of the organizational charts that the firms had been requested to supply with their responses provided some useful information about the actual levels of management or supervision. From these organizational charts and a complementary question about the actual levels of supervision and control it appeared that one third of the 'flat' firms had three and more levels of decision-making, while almost half of the 'hierarchical' firms had only one or two.

Part of this seeming paradox had to do with firm size. Firms with 90 employees can have three levels of authority and still be flat while firms with 10 workers can be hierarchical with only two such levels. Evidence from the case studies suggest that the classification 'flat' or 'hierarchical' entails also an assessment of management style, and of the degree of formality with which these levels of authority and control are enforced.

Another question about work units provided clearer insights into organizational structure. The results (Table 2) show that the majority of the two service sector firms (computer and engineering consultant) are organizing their work in work teams while in telecommunications and wood manufacturing sectors the majority are using a departmental structure. Of the four sectors, only wood firms use production lines to a great extent with almost half of the firms reflecting the classical organization of large batch manufacturing.

Table 2: Style of organizing work by size and sector (percent)

	Departments	Teams	Independent	Production Line	No Answer
Telecommunications	38	33	17	13	
Computers	20	48	20		
Secondary Wood	24	16	12	48	
Offices of Engineers	20	51	26		
Total*	25	39	19	14	4

*percentage does not add up to hundred due to rounding

The various forms of work organization are also often combined. Thus for example, the case studies showed that engineering firms frequently use matrix forms combining vertical and horizontal structures in their project work. Thus 'teams' in this sector tend to be composed differently than in the other sectors.

3.2. External Linkages And Sources Of Information

As mentioned before, small firms rely heavily on external information sources, using networks for sharing information on a regular basis with other firms, industry or related research organizations, and public infrastructure institutions.

3.2.1. Membership In Professional Or Sectoral Associations

Professional and sectoral associations are an important source of information for small firms. Although these associations also provide data and other information to the general public, the gathering and analysis of relevant information serves primarily the membership. Membership in sectoral associations tends to be by firm rather than individual, whereas professional associations have both individual and corporate members.

Table 3: Corporate membership in sectoral or professional association (%)

	Membership	No Membership
Telecommunications	58	42
Computers	64	36
Secondary Wood	68	32
Offices of Engineers	83	17
Total	70	30

Table 3 shows that 70 percent of the survey firms are members of a professional or sectoral organization. Engineering firms are particularly actively engaged in these. This is not surprising given the strong influence of professional engineering associations reflecting a clearly defined professional field of theory and practice. More surprising seems the high degree of membership by wood firms. This is due to two facts. Firstly some of the firms, e.g. specialty saw mills and remanufacturing firms belong to the primary wood associations which in B.C. are very influential and successful in advocating the cause of the forestry industry and in lobbying the government. Secondly, B.C. has actively promoted the creation of specific associations for the secondary wood industry because of the increasing economic importance of the sector and the fact that the interests of primary and secondary wood firms are not identical.

Additional information was requested on the type of information and services firms access from these associations. Of the three main services typically offered by associations, "accessing information and marketing sources" appears to be the most relevant to the high-tech sectors (computer 40 percent, telecommunications 38 percent), whereas "policy related activities" are of special interest to engineering (31 percent) and the secondary wood products sector (24 percent) (Table 4).

"Educational programming and training" activities were on top of the list of secondary wood firms. This is explained by the fact that few of the post-secondary education institutions in B.C. are offering courses and programs of particular relevance to this sector. A new specialized training center (the 'Value-Added Skill Centre for Basic Wood Technology'), was established 1996 by the B.C. Wood Specialties Group (a sectoral association for wood manufacturing firms) to provide basic wood processing skills which are offered at various locations throughout the province. There is a great demand for the programs offered by the Center and the courses are very popular with firms.

Table 4: Use of services provided by professional/sector organizations (ranked no.1 - percent)

	Education and Training	Information and Marketing	Policy Related	No Answer
Telecommunications	8	38	13	
Computers	20	44	4	
Secondary Wood	32	12	24	
Offices of engineers	20	26	31	
Total*	20	29	19	31

* Percentage does not add up to hundred due to rounding

Evidence from the case study discussions indicates that the high-tech sectors are primarily concerned with issues related to keeping their technology and products ahead of the competition, and in strategies related to improving market share. On the other hand, engineering is actively involved in influencing public norms in areas such as waste treatment and environmental legislation: writing standards and serving on advisory boards. This positions them well in terms of responding to requests for proposals, and also makes local companies more attractive as partners in larger projects, where limited resources might require a strategic alliance with a more substantial and out-of-province engineering company. The fact that almost one quarter of wood manufacturers cite policy related activities as their main interest is probably linked to their concerns about the supply and price of raw material that are partly a result of government regulation.

3.2.2. Exchange of Information with Others

As mentioned already, a most important source of information comes from the interaction with other firms. Of the surveyed firms, 82 percent engage in such exchange, most actively the computer firms. This confirms the fluid boundaries of information and knowledge in this sector that has been documented by many other studies (e.g. Saxenian, 1998).

Table 5: Information exchange with other firms by size and sector (percent)

	Information Exchange	No Information Exchange	No answer
Telecommunications	79	17	4
Computers	88	12	
Secondary Wood	72	20	8
Offices of Engineers	86	14	
Total*	82	16	3

* Percentage does not add up to hundred due to rounding

Respondents indicate that the most frequently used source of information are contacts with other firms, both on the vertical axis, i.e. the firm's suppliers and customers, and on the horizontal axis, i.e., with other firms in the same sector or the same geographical area (Table 6). This is consistent with other studies on the particular importance of producer-user relationships. Contacts with supplier and customers are, due to the nature of the relationship, individual, but contacts with the horizontal network are often maintained through industry or sectoral associations.

Table 6: Information exchange with customers, suppliers, and other business partners (ranked no.1 - percent)

	Companies Involved in Information Exchange with				
	Suppliers	Customers	Business Partners	No Answer	other
Telecommunications	21	38	21	4	
Computers	16	44	32		
Secondary Wood	36	28	4	8	
Offices of Engineers	9	34	43		
Total*	19	36	27	3	16

* Percentage does not add up to hundred due to rounding

An exception is the secondary wood sector where firms view information exchange with suppliers as the most important source of information. These firms are especially dependent on the supply of raw material, i.e. fiber, from forestry firms and saw mills. Linkages and information flows are also essential for telecommunications firms, which underlines the necessity for system integrators to be in close contact with the developers and the suppliers of the various elements of their own products.

Engineering firms use other business partners for sources of information and therefore ranked these contacts as a priority. This is explained by the nature of engineering work which requires firms to monitor new technical developments and 'state-of-the-art' applications which require a broad range of contacts and information exchanges. In the case of computer firms, 'business partners' include major hardware and system software developers and providers. It can be assumed that there is a fair amount of overlap between the 'business partner' and the 'supplier' categories.

3.2.3. Customers Feedback During Testing Phase of Products/Processes

Feedback from customers can be either formal or indirect. The flow of information is most intensive during new pilot developments and the testing of prototypes by key customers. This type of testing can be undertaken exclusively by customers or in conjunction with the firm supplying the product, and is quite common in computer services, where 'beta testing' is used to test both hardware and software applications.

Table 7: Testing products using client sites (percent)

	Testing Products at Clients' Sites	
	Yes	No
Telecommunications	71	29
Computers	88	12
Secondary Wood	36	64
Offices of Engineers	34	66
Total	55	45

Table 7 indicates the frequency of testing at the site of, and by clients. More than half of all firms are using this type of testing to refine prototypes or pilot applications in order to save costs and time. Once a product enters the market, defects can be dangerous and expensive if customers opt to sue or boycott goods. Having a company test a product can save months in development time back in the lab. This type of feedback from clients is particularly frequent in the high-tech industries. For example, 88 percent of the computer consulting firms and 71 percent of the telecommunications firms responded that they were using client testing for feedback. For engineering firms, customer testing of prototypes and pilot applications of new technical processes and their feedback are essential for eventual re-design and the fine-tuning of controls.

Testing of products in close connection with clients has important implications for the internal organization of a firm, as contacts with clients are no longer confined to senior management and the marketing and sales units. As production units are becoming directly involved in direct customer contacts, the flow of communication can no longer be vertical but must be organized in a horizontal mode. This trend is well documented in the case studies and has not only repercussions for work organization but also the role of clients and the market in the process of innovation, which is the theme of the following section.

3.3. The Innovation Process

Innovations are generated by a complex process often involving many players and dependent on a multitude of factors, both internal and external to the firm. Evidence from the case studies as well as many other studies suggests that although innovations often share some common features, they have specific trajectories and therefore are quite distinct - which is the reason why there is no general 'best practice' model of innovation that could easily be emulated by others.

This section is concerned with the 'style' of product or process innovation and the use of external and internal sources of information.

3.3.1. Style of Innovation

The case studies demonstrated that there are several ways of arriving at a successful innovation. In some cases systematic planning and R&D formed the basis of innovation, while other firms relied on a non-linear and serendipitous process. However as explained above, the process is frequently both linear and planned, and somewhat chaotic and haphazard in nature.

When asked whether their product or process development takes place according to a definite plan, schedule, or pace, or was more serendipitous, 57 percent of the survey firms responded that development was planned and systematic, whereas more than one fifth said the process was better characterized as serendipitous (Table 8).

Table 8: Style of innovation (percent)

	Systematic	Serendipitous	Other	No Answer
Telecommunications	83	13		
Computers	84	8	4	
Secondary Wood	44	32		
Offices of Engineers	29	29	6	
Total	57	21	3	19

Firms from telecommunications and computer services provided the most consistent and identical response rate, with 84 percent maintaining that development was planned and systematic. By contrast, engineering firms were split in the middle on this question. However more than one third of engineering firms did not answer this question which suggests that neither of the two suggested styles appropriately described the dynamics of the process. This is in line with evidence from the case studies from this sector which show that engineering innovations were made as part of a specific contract, or upon the demand of a particular client, rather than in the form of a linear and long-planned development.

3.3.2. Origin of Innovation From Inside The Firm

Innovation was defined earlier as a complex process involving typically a variety of 'players'. Responses to the survey question "Where in the firm does innovation typically originate from?", confirm this concept of a collective activity, showing that in 57 percent of the firms, ideas and concepts come from several people (Table 9). The computer sector sticks out again as the most team-based.

Table 9: Origin of innovative ideas (percent)

	Individual	Department	Project Team	Senior Management	Mixture	No Answer
Telecommunications	25		4	13	58	
Computers	16			12	68	
Secondary Wood	12	4		24	52	
Offices of Engineers	14		11	20	51	
Total*	17	1	5	17	57	4

* Percentage does not add up to hundred due to rounding

Firms attributed 16 percent of innovations to individuals, with the greatest concentration in telecommunications companies. Innovations originating from the top, i.e. senior management, accounted for 17 percent of responses, with wood firms, and engineering firms accounting for the largest share. In the wood sector this can be related to the traditional hierarchical structure and top down management style. Engineering firms owners are senior professionals who are centrally involved in the definition and acquisition of new projects.

Thus, survey findings confirm the information provided by the case studies in that, while most innovations tend to be the result of a team effort, there is a definite role for individual staff members with creative ideas. In the high-tech sectors of telecommunications and computers, these individual innovators clearly outweigh the role of senior management. In engineering offices, innovative ideas and concepts tend to come from both individuals or project teams, accounting together for 25 percent of firms.

The survey results cannot explain exactly how the innovative ideas from the various actors were communicated and put to use by the firms. Probably the trajectories of these innovations vary widely, but case study evidence shows that the prerequisites for organizational learning that were mentioned above, namely individual or team autonomy, unhindered flow of information, flexible and flat structures, and an active learning strategy, were all found in innovative firms. The fifth factor, appropriate skill levels, will be discussed below.

3.3.3. External Sources of Information

The question probing external sources of information is closely linked to the query into how small firms gain access to information generally (discussed above). In this question, firms were asked about the usefulness of a number of information sources for innovative development.

Table 10 indicates that overall trade shows and related activities are the most frequently cited. Such trade shows, fairs or exhibitions typically include industry presentations, demonstrations or workshops, and firms get the opportunity to meet not only suppliers of new equipment, systems, software applications and other related solutions, but also competitors, and present or potential customers. Thus they constitute a most important marketplace of ideas, information and business contacts.

Table 10: Usefulness of various sources of information (frequency of response - percent)

	Academic Conferences	Internet	Universities/ Labs	Publications	Professional Seminars	Trade Shows
Telecommunications	38	58	29	67	54	75
Computers	44	84	48	84	80	80
Secondary Wood	24	28	16	52	52	84
Offices of Engineers	57	49	46	80	80	69
Total	42	55	36	73	68	76

'Publications', ranging from scholarly publications to professional and technical periodicals or trade journals of various kinds, were rated to be of great value generally, but especially to computer and engineering firms which employ most workers with a higher education background. This reliance on publications may change however. The Internet was still in its infancy when the survey was conducted and it can probably safely assumed that its importance has disproportionally increased since that time. In fact, for the purpose of information retrieval, it may have well surpassed all other sources of information due to its easy accessibility and actuality.

3.4. Human Resources Development, and Individual and Organizational Learning

The type of knowledge and the competencies that firms need in order to successfully innovate are related to the skills level of the workforce, the firm's human resources strategy, and, most importantly, collective and individual learning.

Recent literacy surveys show that the years of school attendance are not always a reliable measure for assessing literacy levels (OECD 1995c and 1997). Nevertheless, studies on adult learning have consistently found that the individual's learning is closely linked to previous educational attainment and experience. Educational attainment is thus used as a proxy for the ability and the motivation to engage in further learning. Table 11 shows the educational attainment of the workforce in the four sectors.

Table 11. Educational attainment of employees by industry sector (percent)

	No High School Completion	High School Diploma	Degree	Post-Graduate Degree	Other Qualifications
Telecommunications	7	39	28	5	21
Computer	13	33	42	6	6
Secondary Wood	13	68	6	0	12
Office of Engineers	4	32	37	17	11

Workers without high school diplomas are found in all four sectors. However, they are relatively frequent in the secondary wood industry and, somewhat surprising, in computer services. Workers with high school diplomas but no further qualifications represent roughly one third of the workforce in three sectors, but not surprisingly more than two thirds in the wood industry, which has a tradition of hiring unskilled workers.

In the two service industries, computer and engineering consulting, around half of the employees have university degrees. The result is a polarization of skill levels in both industries which is a potential problem for learning processes that involve the organization as a whole. The wood sector has a very low level of degree level skills reflecting the industry's traditional occupational and organization structure.

The category 'others' reveals the other types of qualifications found in the workforce. Primarily this group consists of people with technician qualifications from either the B.C. Institute of Technology, a Community College, or one of the numerous private vocational schools for technical training. As the case studies revealed, the share of employees with technical diplomas is particularly high in engineering offices, where computer assisted design and geographic information systems are becoming an integral part of operations. In the wood industry such 'other qualifications' refer mostly to journeyman diplomas (the gender-specific term is justified here as the great majority of skilled workers in this industry are male).

3.4.1. Formal Training

As mentioned before, the incidence of training in small firms is lower than that in larger companies. Nevertheless, more than three quarters of firms responded affirmatively when asked training was part of the firm's business plan. Yet less than one third had a training budget and tracked their training expenses. Almost two thirds of training is provided in-house, approximately one quarter by professional associations or private institutions, and only seven percent by public institutions.

Asked about the barriers to provide training for their employees, the majority of companies cited time, i.e. conflict with work, and training expenses as the most important obstacles. Working time lost during training, which is a factor well known from studies from other jurisdictions, was a concern especially of the two sectors that tend to organize their work in teams, computer and engineers, whereas firms in the two manufacturing sectors seem to be able to more easily replace workers on the production line while they are away for training. However, this problem of replacing workers at work is a problem which is more easily related to the firms' size than other obstacles such as cost or lack of suitable training courses.

3.4.2. Informal Learning Activities

The problem of replacing workers who are away for training is the main reason why small firms prefer on-the-job training and other informal types of training or learning which takes place on site and make learning part of the work process. If such informal training is to be more than learning-by-doing, arrangements must be made for tutoring or mentoring by experienced workers, cross- training by co-workers, or for some kind of assistance for independent learning. Judging from case study evidence, such arrangements are usually made if the management has a learning strategy, even if it is not explicit.

Learning strategies that include informal ways of acquiring know-how can encompass various activities. The case studies show that the most frequently cited activities include regularly scheduled brainstorming meetings, the sharing of conference papers, reporting on seminars and trade shows, mentoring situation of various kinds, and work-related social activities which include the sharing of work-relevant information and experience.

When survey firms were asked about the use of such strategies (Table 12), the use of informal mentoring appears to be the most common way of sharing information and passing on knowledge. This approach is well suited to 'breaking-in' new employees with no or little work experience, and important in small firms because of their tendency to hire new graduates rather than workers with previous experience. Of the five strategies or techniques mentioned in the questionnaire, this category is also the most varied. It ranges from teamwork with an experienced team leader as a mentor to a one-to-one pairing of a more experiences employee with somebody more junior, or new to the firm. Larger firms appear to be using this strategy more regularly than their smaller counterparts. It is also the most frequently cited informal learning activity in engineering firms where project team organization is particularly conducive to this form of imparting know-how.

Table 12: Informal learning strategies for sharing information, knowledge, and experience (frequency of response - percent)

	Sharing Papers	Brown Bag/'Lunch & Learn' Sessions	Brain-Storming	Informal Mentoring	Retreats/Social Activities	Other
Telecommunications	29	17	67	71	8	
Computers	40	36	80	68	24	4
Secondary Wood	8	36	44	44	8	4
Office of Engineers	63	71	60	80	20	
Total	38	43	62	67	16	2

Almost two thirds of firms use 'brainstorming', a term that can describe a variety of forms. Brainstorming is typical for the computer service sector where this form of sharing of information and experience in the joint creation of perspective and knowledge has become the most common form of collective learning. Attending combined lunch-and-learn sessions ranked third overall in collective learning activities. This form is frequently used by engineering firms who also frequently share conference papers.

Only a minority of firms uses work-related social activities outside the work place such as retreats as an appropriate instrument of the firm's learning strategy. An exception are some computer and engineering firms where such social activities are organized for the purpose of exchanging information and, at the same time, promoting a sense of community and firm identity.

As mentioned before, all of these approaches to systematic but informal learning through sharing information and knowledge, are conditioned by factors such as a flexible and flat organization, a shared body of 'core knowledge', and an active learning environment. The level of educational attainment, and the occupational 'habit' or 'ethos' of different groups of personnel within the company also play role in the ability and willingness to learn and to contribute to the learning of colleagues.

Probably the most significant factor is the content of the work itself, and the extent to which it requires or encourages learning (Rubenson & Schuetze 1996). Contrary to the widespread perception, this is not necessarily predetermined by the sector or by the degree of sophistication of the technology used. The space and the incentive the workplaces give for learning depends to a great extent on the strategic choices firms make with regard to the way they organize work and actively embrace human resource development.

4. SUMMARY

In the first part of the chapter, five themes and concepts were introduced describing different approaches to the knowledge management process in small firms. Their common focuses are the structures and processes that enable an enterprise to harness different resources within and from outside the firm to use existing knowledge and to create new knowledge. Except for some small companies that use highly specialized technology or are science-based and in which the principal sources of know-how reside within the organization, most small firms must look outside for some of the information they need to be successful in the market. The challenge for these firms is partly different from knowledge management in large firms where the necessary information and know-how consists of finding the experts and expertise within the organization. Linkages with their business environment and networks play therefore a central role in small business knowledge management.

In order to use these linkages and networks effectively as sources of information and knowledge, firms must also have 'absorptive capacity', i.e. the ability to identify the information that are relevant for the firm, and to put them to use. Work organization, access to information, communication, and formal and informal learning are factors that influence this ability. Because of their small size, i.e. the geographical proximity of the various units, the familiarity of the staff with each other, and the greater informality of personal relations, information flows within the firm and communication between the staff members are in principle easier to manage than in larger companies, even if barriers of status or expert language may be a factor.

In the second part, results from a study on innovation, knowledge and human resources management in small firms are used to show how these concepts manifest themselves in the practice of small companies. The analysis concentrates on four areas which are relevant to knowledge management: firm organization, linkages with the business environment and external sources of information, the innovation process, and human resources development and individual and organizational learning. The analysis shows many similarities between firms that can be related to their similarity in size, even if there are some significant differences due to different industry and professional cultures in the four sectors.

Chapter Seven
MANAGING SURVEYS ON TECHNOLOGICAL KNOWLEDGE:
The French Experience in the Nineties

Stéphane Lhuillery[1.]
C.E.P.N., C.N.R.S., Paris

"Measure what is measurable, and make measurable what is not so."
Galileo (1564-1642)

"Measurement of a thing is a prerequisite to managing it"
John Von Neumann (1903-1957)

1. INTRODUCTION

At the level of the enterprise, the knowledge capital of firms is presumed to be the dominant source of competitive advantage. When this issue was first recognized, attention was focused on quantifying the production of new knowledge obtained through the employment of new technology. At the beginning of the 1970's, a first wave of data collection on research and experimental development (R&D) activity was introduced by the Organisation for Economic Co-operation and Development (OECD). A second wave during the 1990's dealt with technological innovation. This latter initiative led to the development of a number of new indicators at both the input as well as at the output level. These indicators can be summarized as follows. On the input side, the innovation cost indicator is now available (see Eurostat, 1997). It includes more than the usual bulk of R&D expenditures and better fits the resource allocation for innovation. International comparisons appear to show that this measurement indicator is quite reliable since the share of innovation costs is quite homogenous among the European Community (E.C.) countries (Eurostat, 1997, 1998). However, though firms are aware of how many dollars they invest in the innovation process, it seems that they are not able to precisely map the different kinds of resources that are involved (see Evangelista et al., 1994 on CIS1;

[1.] I gratefully acknowledge those people at the Ministry of Education (DPDC3), of Industry (SESSI) and Nathalie Greenan for helpful information. Many thanks to Louise Earl from Statistic Canada for her assistance. Opinions and errors are the author's alone.

Kleinknecht, 1997, Francoz, 2000, on CIS2)[2]. The question of innovative inputs therefore remains murky and further investigations are still required. On the output side, the sale of innovative products are the best indicator available thanks, to the CIS initiative (see Kleinknecht and Wunderink, 1999). Even if the breakdown by degree of innovation does not seem very reliable, the global bulk of product innovation seems to be known by enterprises, even multi-project ones. However, it is rarely mentioned that this indicator provides more than just a measurement of technological capabilities of firms, it also covers complementary know-how: new marketing methods, new organization design, new information systems, new product design, etc...

As pointed out in a famous French dictionary (Littré, 1863), "innovation" is at the same time the result of a process and the process itself. As already stated, the first aspect of innovation as a result of a process has been extensively considered. The second aspect, innovation as the process itself has been scantily investigated. Indicators on the means of information and knowledge transfers (in CIS1) or even the items on the partners of R&D cooperation (in CIS1 and CIS2) provide clues into the absorptive capacities and practices implemented at the firm level. Internal processes, however, often have been neglected in representative surveys despite the growing literature on knowledge management or knowledge economics[3]. A third wave of surveys is thus required to collect information on practices implemented to produce and manage knowledge. Although such an inquiry is based on as yet "primitive" view of knowledge production and management (Teece, 1998), it can be founded on past experiences in collecting R&D and innovation statistics.

The purpose of this chapter is to provide insight into the French questionnaire design, data collection and analytical experiences during the nineties. The chapter proceeds as follows: Section 2 reviews the different indicators that have been developed to specify more precisely the innovation function. This section does not document the advantages and weaknesses of the usual spread of new input or output indicators on innovation that are well chronicled elsewhere (see Kleinknecht, 1991; Smith, 1999; Kleinknecht, 2001). It focuses on efforts made to consolidate and accumulate new results from the R&D inputs, then introduces the new complementary R&D indicators, and finally provides an overview of the non-technological indicators of innovation outputs. Section 3 leaves the innovation function point of view to focus on the questionnaires dedicated to innovation management practices. Mainly focused on new technological knowledge, due to the original statistical inquiry on competencies, this section also includes lessons from the surveys conducted on innovation in organizational practices.

---

2. Although the French CIS2 results are consistent with the German CIS1 results (Licht et al., 1995)

3. The "PACE" questionnaire on appropriation (Called also Yale2 survey) includes a question on the strength of internal interactions between R&D, production and the marketing functions.

We show that a new breed of questionnaire could be launched investigating the management practices at the firm level as well as at the employee level. Conclusions are presented in the final section.

2. DEEPENING & EXTENDING THE INNOVATION FUNCTION: THE INPUT-OUTPUT VIEW

At the input level, R&D is falsely considered as a standard and steady activity. Recent developments give interesting results for the understanding of innovative activities. At the output level, the first French innovation survey ("EAE 1991" hereafter) and the recent French survey on competencies ("Competencies" hereafter) deal with technological innovations as well as non-technological innovations. (see Table 6 in Appendix 1) The link between the different types of innovation is then described.

2.1. R&D Indicators

Although the new qualitative data on innovation costs are interesting, the CIS survey and the traditional R&D survey provide four important aspects of the R&D management and measurement. The usual definition introduces a threshold among the R&D performers since the R&D "execution" must be "systematic" (OECD, 1993a). However many small and medium-sized enterprises (SMEs) or services firms are still able to keep their R&D informal. Their informal R&D activity includes less than one full-time equivalent resource per year or a non-specialized unit. The uptake of this organizational practice in different countries is now well documented (see Kleinknecht and Reijnen, 1991; Santarelli and Sterlaccini, 1990; Lhuillery and Templé, 1994). Also, it has led to several modifications of R&D and innovation questionnaires.

On the second CIS questionnaire a new question was introduced: "Did your enterprise engage in R&D between 1994 and 1996?" The possible answers were "Continuously", "Occasionally", or "Not at all". 44 percent of the French innovative firms declared a continuous R&D activity and 32 percent reported discontinuous activity in CIS2. The ratios are 64 percent and 26 percent for large firms respectively (500 employees and more). At the same time, the concept was investigated through a new question introduced in the annual R&D questionnaire that identified the existence of a dedicated R&D department or center. Half of the R&D firms declared such formal R&D organization either in manufacturing, industry or service sectors (in 1996).

A related limit of the Frascati data is the orientation toward R&D "execution": firms with no internal execution are not included in the R&D surveys even if they contract out R&D to an external body. The CIS survey provides evidence that 6 percent of innovative firms with a positive R&D budget therefore are missed by standard R&D investigations. Even if the occasional R&D performers or pure R&D contract performers have no significant impact on the bulk of the R&D expenditures, it is useful and important to delineate R&D firms that are the more likely to produce S&T knowledge.

A third interesting difficulty of the R&D data is the quality of the usual distinction between fundamental research, applied research and development. Even if several enterprises tend to make a distinction among the R&D activities, the different levels of R&D are barely the same each year. Although the level of R&D expenditure has increased over the last 20 years, the share of fundamental or even applied research in the R&D budget is quite constant (4 percent of the internal R&D expenditure of firms). The result is consistent with the U.S. figures (NSF, 2000). However, the result is at odds with the usual assertion that firms are getting closer to basic research in order to be able to participate in scientific networks (Rosenberg, 1990; Pavitt, 1988). Further analysis suggests that the meaning of "fundamental" is not the same in the pharmaceutical industry as in traditional manufacturing industries. The regularity of the results over the last 20 years also tends to show that firms are not able to understand and/or distinguish between the three amounts. This weakness is a serious limitation to investigate since R&D data on the degree of openness of firms reinforces the relative importance of alternative indicators such as R&D cooperation or academic publications.

A fourth weakness of R&D indicators that has been pointed out, is that a firm can execute R&D in fields other than those in which they produce. The French R&D survey uses a dual classification: one reports the main sector of activities; the other reports for multi-branch R&D activities which are supposed to be completed on separate questionnaires for each R&D branch. Of course, such a scheme is not attractive to respondents and only 21 out of 3363 of them completed several 1996 questionnaires. Although the sector-branch matrix is very diagonal in manufacturing industries, the double classification is useful in dealing with the growing number of R&D spin-offs that are classified into R&D services (see OECD, 2000b). A lack of a dual classification can lead to an underestimated R&D investment of some manufacturing sectors, such as the automobile industry in France. Beyond the R&D bulk accountancy, branch classified data would also be useful in computing the technological proximity of firms. This measure is usually computed through the market flows of intermediate or final goods or patent classification (à la Jaffe, 1986) or the proximity of firms within the scientific fields (see Cohen and Levinthal, 1989, on the topic). The introduction in the new questionnaire on researchers of a question on the field of research (Biotech, Software, Environment, New materials, Social Sciences), although partial, is a first step into a recognition of the complexity and complementarities among technological knowledge produced by firms.

2.2. Managing Innovation Employees

A second possible means to deal with knowledge management is to focus on human capital management by firms. Even if the possibility is recognized and encouraged at the international level (see "The Canberra manual" from the OECD, 1995a) few serious investigations are available to date. A special annual survey on researchers and an annual survey on occupational structure have been available for many years in France. Until now, these two sources did not receive the attention that they deserve.

Researchers' population is studied by a non-mandatory, annual, French survey ("Researcher" survey hereafter). The set of questions deals with their degrees and even with their origin and destination for arrival and departure in the R&D activities. Lacourrège and Jacquin (2000) give evidence of the slow rise of the share of Ph.D. graduates in R&D enterprises (about 15 percent in 1997). The results suggest that industrial R&D is getting closer to scientific, academic research corroborating the bibliometric works on scientific publications (Hicks, 1995) or patents citation (Hicks et al., 2001). Such data also allow us to distinguish the industries or branches with different uses of the "internal market": in some cases, a large part of researchers are "home made" by internal promotion and do not hold university degrees. In other industries, a large part of hired researchers are university graduates (Bachelor of Science, Masters, Ph.D.).

As Graph 1 shows, the distinction does not cross the "High-Tech/ Low-Tech" classification. Some Low-Tech R&D branches hire mainly high level graduates and have a low level of internal promotion (e.g. Consultancy) whereas other Low-Tech R&D branches rely on non-market or professional training (e.g. Textile, Transport). While the gap is reduced as the R&D intensity increases, a High-Tech R&D branch such as Aeronautics uses non-graduates more than the average does (represented by the two axes) or the pharmaceutical industry. Such results are interesting because they emphasize the importance of internal training and learning between the different branches of R&D. They can also be taken as a signal of the share of tacit knowledge in the R&D branch or sectors that cannot be found on the labor market. Another interesting result is that the share of internally trained researchers has remained relatively stable over the 10 last years (e.g. 18 percent in the automobile branch in 1991).

Graph 1: R&D Intensity and the Share of Internal Promotion in R&D (firms, by R&D branch (in 1998)

Source : Bureau des études statistiques sur la recherche - DPD-C3, Ministère de l'Education

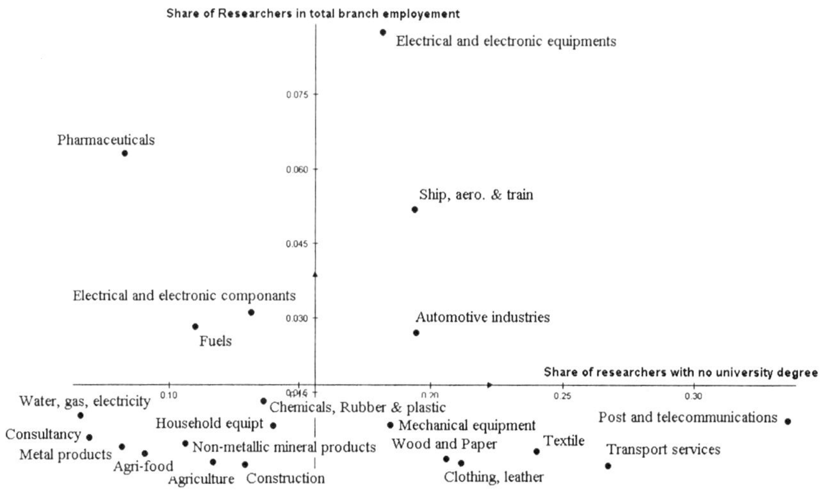

The survey on researchers is original but furnishes only a restricted view of the innovative process. A broader account is available through an in-depth French survey at the plant level on the occupational structure (the "Employment" survey hereafter). This annual and mandatory survey provides a possible measurement of innovation employment and its evolution[4]. The management of knowledge is thus considered here through the management of human capital. An interesting exploration is the consequence of human capital structure and its evolution on the innovative capacities. For example, employees can be disaggregated into very distinct classes as shown in the Table 1 restricted here to white-collar workers:

4. Educational statistics provide an alternate way to grasp information on the knowledge structure of firms. Such indicators can be build at an aggregated level (see, Hansen, 1999 on Canadian Census data) or at the firm level (Jacobsson & Oskarsson,1995).

Table 1: A suggested classification into 12 classes of white-collar workers (out of 353 possible positions)

N°	Classification	Title
1	23	Business Executive
	37	Administrative and sales managers
	371	Management and general administration.
2	3721	Managers of economic, financial and sales studies.
3	3722	Managers specializing in recruitment or training.
4	3723	Organizational managers or administrative and financial comptrollers.
5	3728	Documentation or archiving managers
6	3722 to 3727	Other management and administration managers
	382	Research, studies, testing, electronic data processing
7	3828	Electronic data processing engineers and managers (excluding technical sales)
8	3821 to 3827, 3829	Other studies or research engineers and managers
9	383	Manufacturing engineers and managers
	384	Maintenance, new construction, related production functions
10	3842 (only)	Industrial purchasing and supply engineers and managers
11	373	Sales, publicity
12	385	Technical sales engineers and managers (professional or intermediate equipment)

Source: Dares, 2000[5]

Such a classification allows us to capture the structure of qualification that is potentially involved in the innovation process. Also, it introduces the knowledge required to commercialize innovation; accounts for production level knowledge; identifies workers involved in design activities; presents discontinuous R&D; introduces the role of computers in developing the required knowledge and specifies different absorptive capacities through documentation as well as material and machine supplies.

To date, very few investigations are available using this data set. One possible route of investigation is to focus on the impact of the occupational

---

5. Direction de l'Animation de la Recherche, des Etudes et des Statistiques, Ministère de l'emploi et de la solidarité - secteur emploi, France [Research, Studies and Statistics Management Branch, Department of Employment and Solidarity - Employment Sector, France].

structure on the firm's productivity. On this topic, Duguet & Greenan (1998) found that innovation is skill-biased that is, it leads to a firm employing a higher share of white-collar workers. Greenan et al. (2001) built a more detailed variable set including four categories of general skills[6]: design versus execution tasks crossed with administration and production activities. The results suggest that the share of administrative managers (e.g. the aggregation of the defined classes number 3721, 3722, 3723...) is positively and significantly related to the firm's productivity.

The present efforts made on developing the innovation input indicators are also achieved for innovation outputs as it is shown by the two following sub-sections.

2.3. Including Non-Technological Knowledge

The CIS innovation surveys do not aim at capturing the non-technological side of innovation activities (OECD, 1996b). Nevertheless, the importance of this kind of innovation is acknowledged (see Annex 2 of the Oslo Manual and the last question of the main CIS3 questionnaire or even Schumpeter, 1911). Two surveys dealing with technological innovation introduced questions on non-technological innovations. A first attempt was made by the first innovation survey (non-mandatory) launched in France by SESSI (EAE 1991) investigated packaging innovation[7], organizational innovation[8] and marketing innovation[9]. Such complementary views on innovation are also investigated in the first surveys on competencies for innovation (Competencies) launched by SESSI in 1997 (see Appendix 2).

Table 2: First French Innovation Survey (19589 firms)

Innovation	Product	Process	Technological	Packaging	Organization	Marketing	Non Technological	All types	At Least One
Column	(1)	(2)	(1)È(2)	(3)	(4)	(5)	(3)	È(4)	È(5)
Large	0.82	0.78	0.88	0.28	0.38	0.20	0.54	0.05	0.89
SME	0.53	0.44	0.60	0.14	0.18	0.12	0.29	0.01	0.61
All	0.54	0.46	0.61	0.15	0.19	0.12	0.30	0.01	0.63

EAE 1991 / Period: 1986-1990

6. The already mentioned commercial or marketing positions are thus not sustained in the paper.
7. "Creation or substantial improvement from a technological point of view the packaging modes".
8. "Introduction of substantial organizational innovation linked to technological innovation".
9. "Introduction of substantial commercial innovation".

Table 3: Technological and non-technological innovation rate, by class of size (22358 manufacturing firms)

Innovation	Product	Process	Technological	Design	Marketing	Non Technological	All types	At least one
Column	(1)	(2)	(1)	È(2)	(3)	(4)	(3)	È(4)
Large	0.52	0.47	0.57	0.43	0.32	0.52	0.16	0.70
SME	0.38	0.34	0.49	0.26	0.21	0.36	0.06	0.58
All	0.39	0.35	0.49	0.27	0.22	0.37	0.07	0.59

Competencies survey / period 1991-1993

Although it is hard to compare the results from the EAE 91 and the Competencies survey because the periods considered are different (respectively covering 5 and 3 years) both show that non-technological innovation is often introduced by manufacturing firms. Design innovation and marketing innovation are less frequently addressed than technological innovation but both are a concern for more than one third of manufacturing firms (in Table 3). The innovating firms, taking into account differences in firm size, are thus 10 percent more numerous and represent 59 percent of the weighted sample. The results are consistent with the 69 percent found in German firms from the CIS2 where a distinction between "technological" or other forms of innovation was not introduced (Archibugi & Sirrili, 2000), taking into account the differences of firm size between the two countries[10.]. Large firms are more likely to innovate from a technological point of view but are also more frequently design or marketing innovators.

Another argument to put forward is that the Competencies survey does not encompass all the kinds of innovation included in the German CIS2. The EAE91 survey suggests that organizational innovations, even restricted to those linked to technological innovation, are at least as frequent as marketing innovations.

The 19 percent organizational innovation rate is quite low when it is compared to the results of recent surveys dedicated to organizational innovation. The survey on the organizational change at the production level (the "COP" survey hereafter, see Table 7, Appendix 1) reveals that 79 percent of firms of more than 49 employees change their production organization (Greenan, 1995). The rate is higher than the 66 percent rate found for Swedish firms (NUTEK, 1966) covering a 10-year period. On a restricted period of 3 years, organizational innovation is declared by 52 percent of Danish firms

---

10. Another argument is that the breakdown into marketing and design reduces the percentage of innovative firms. This effect is clearly demonstrated when the distinction between the degree of innovation of the introduced product is investigated in CIS2. Compared to the Competencies survey with the same method, period and response rate, the product innovation rate is 34 percent instead of 39 percent.

(Gjerding, 1996), and 49 percent of French firms (on a second survey called "Response" from the DARES, Coutrot & Paraire, 1996, and Table 7, in Appendix 1).

The size effect that is usually found for technological innovation is also obtained for non-technological innovation (see Table 2 and Table 3). The size effect on organizational innovation seems very robust and found by surveys on organization in France, Germany, Sweden and Denmark (see respectively Greenan, 1995, Kinkel et al., 1997, NUTEK, 1996, Gjerding, 1996).

Table 4: Technological and non-technological innovation, by sectors (EAE 1991)

NAP	Label	Product	Process	Techno.	rk	Packaging	Organization	Marketing	NonTechno.	rk	All types	At least one	rk
(1)	(2)	(1)È(2)			(3)		(4)	(5)		(3)È(4)È(5)			
U02	Food	0.61	0.46	0.66	9	0.36	0.17	0.15	0.44	2	0.02	0.69	9
U03	Energy	0.33	0.39	0.44	16	0.15	0.19	0.15	0.27	13	0.05	0.47	16
T7+ T8+ T13	Other mining, metal prod.	0.45	0.54	0.63	11	0.12	0.23	0.13	0.31	9	0.01	0.64	12
T09+ T10	Glass, Construction mat.	0.55	0.46	0.63	12	0.11	0.18	0.11	0.27	11	0.01	0.65	11
T11	Basic Chem., synthetic fibers	0.74	0.68	0.80	2	0.17	0.17	0.11	0.31	7	0.01	0.80	2
T12	Pharmac. & fine chemicals	0.81	0.58	0.84	1	0.38	0.25	0.19	0.52	1	0.03	0.84	1
T14	Mechanic suppliers	0.66	0.50	0.70	7	0.10	0.21	0.14	0.30	10	0.01	0.71	7
T15A	Electric. & Electro. Supp.	0.72	0.54	0.76	3	0.11	0.23	0.14	0.31	8	0.02	0.76	6
T15B	Household equip, Ind.	0.74	0.55	0.76	4	0.17	0.28	0.14	0.41	3	0.01	0.77	3
T16	Automobile & Train	0.70	0.60	0.76	5	0.14	0.24	0.14	0.34	6	0.02	0.76	5
T17	Ships, Aircraft	0.59	0.55	0.66	10	0.05	0.21	0.09	0.25	15	0.01	0.66	10
T18	Textile, Clothing	0.31	0.26	0.36	17	0.08	0.11	0.07	0.17	17	0.01	0.36	17
T19	Leather, Shoes	0.44	0.37	0.50	14	0.09	0.14	0.09	0.21	16	0.01	0.52	14
T20	Wood, furnitures	0.50	0.41	0.57	13	0.14	0.15	0.12	0.27	12	0.01	0.58	13
T21	Paper prod.	0.62	0.50	0.67	8	0.25	0.17	0.13	0.36	5	0.02	0.70	8
T22	Print, Edition	0.37	0.37	0.46	15	0.10	0.17	0.11	0.26	14	0.00	0.49	15
T23	Rubber, Plastics	0.70	0.57	0.76	6	0.20	0.24	0.16	0.38	4	0.02	0.76	4
	All	0.54	0.46	0.61		0.15	0.19	0.12	0.30		0.01	0.63	

The heterogeneity of the expected technological innovation rate, after accounting for differences in sectors, is also apparent when non-technological innovation is introduced (see Table 4). While packaging innovation is very important in food industries, it is also important to pharmaceuticals or paper and printing industries. Organization and marketing innovations are similarly important to the printing and publishing, energy or metal products industries, whereas the aircraft and basic chemicals industries depend on technological knowledge. The other industries rely upon the last three types of innovation: technological knowledge completed by new organizational practices and new marketing techniques.

Table 5: Technological and non-technological innovation rate, by NACE sectors (Competencies survey)

Innovation:		Product	Process	Techno.	Rk	Design	Rk	Marketing	Rk	Non Techno.	Rk	All types	At least	Rk
(NAF36)	Column Sectors	(1)		(2)		(1)È(2)				(3)		(3)È(4)		(4)
C1	Clothing, leather and footwear	0.15	0.18	0.24	14	0.27	11	0.18	13	0.32	11	0.03	0.41	14
C2	Printing and publishing	0.18	0.27	0.33	13	0.19	13	0.17	14	0.27	14	0.03	0.43	13
C3	Pharmac., fragrances	0.64	0.44	0.68	2	0.43	2	0.35	2	0.53	2	0.14	0.81	1
C4	Household equipment industries	0.41	0.32	0.47	11	0.52	1	0.30	3	0.57	1	0.12	0.66	7
D0	Motor Vehicles industries	0.6	0.44	0.65	4	0.38	4	0.22	7	0.46	6	0.09	0.73	4
E1	Train, Ships & Aero	0.53	0.41	0.63	6	0.34	5	0.37	1	0.49	3	0.13	0.78	3
E2	Mechanical Equipt	0.42	0.32	0.51	7	0.22	12	0.20	11	0.32	12	0.06	0.57	10
E3	Electric. & Electro. Supplies	0.65	0.49	0.74	1	0.39	3	0.20	10	0.47	4	0.08	0.79	2
F1	Mining, Glass, Construc. Prod.	0.36	0.33	0.45	12	0.28	9	0.21	9	0.35	10	0.08	0.55	12
F2	Textiles industries	0.38	0.36	0.49	9	0.3	7	0.24	6	0.39	8	0.07	0.64	8
F3	Wood and Paper	0.39	0.32	0.47	10	0.28	8	0.26	4	0.41	7	0.09	0.58	9
F4	Chem., Rubber & plastic prod	0.59	0.42	0.64	5	0.27	10	0.22	8	0.38	9	0.07	0.71	6
F5	Basic metal, metal products	0.31	0.38	0.49	8	0.17	14	0.20	13	0.29	13	0.05	0.55	11
F6	Electric. & Electro components	0.58	0.44	0.66	3	0.33	6	0.25	5	0.46	5	0.09	0.72	5
	All	0.39	0.35	0.49		0.27		0.22		0.37		0.07	0.59	

Calculated by the author - (Weighted sample = 22358 manufacturing firms)

A direct comparison with the results of the Competencies survey is not possible due to use of a different classification system and the introduction of design innovation to the detriment of organizational innovation. Table 5 suggests that household equipment as well as clothing, leather and footwear industries, as expected, depend on design innovations and to a lesser degree on technological innovation. Both kinds of innovation are required in the automobile industries, electric and electronic suppliers, automotive or chemicals, plastics and rubber industries even if the design side is not as important. Finally, technological, design and marketing innovation are central innovation agents of pharmaceuticals as well as transportation industries (train, ship, and aircraft).

2.4. Interdependent Polymorphic Knowledge

The results suggest that technological innovation is central to innovative firms' activities. However, this result does not imply that innovation is impossible without an important contribution of technological knowledge. The food industry is based on new recipes and packaging even if one can argue that artificial flavors or new processes are also implemented in this industry. It also suggests a positive correlation between technological innovation and non-technological innovation or at least correlation between the different kinds of knowledge produced within a firm. Even if a general investigation on the complementarities among all the different types of innovation is not yet available, complementarities between organizational innovation and technological innovation were extensively explored in France.

Using the EAE 91 survey, Greenan and Guellec (1998), show that radical technological innovators are more frequent organizational innovators. The COP survey clearly shows that new technologies induce better communication between the production function and the other functions in the firms (Greenan, 1995; see also Greenan and Guellec, 1998 on other data). Other surveys on organizational innovation corroborate these findings (see Gjerding, 1996; Lay, 1997) even if correlation between organization and process innovation seems weaker (NUTEK, 1996).

The causality between the two kinds of innovation remains to be analyzed. Greenan (1995, 1996) suggests that the causality can be the other way around: the choice of a flexible organization induces the adoption of new technological processes even if the link is slight for higher levels of process innovation (see also NUTEK, 1996).

Further investigations should include a survey of the different kinds of knowledge employed and produced at the firm level. It would be interesting to follow this up and to match a complete view of innovation inputs with a widened measurement of innovation outputs. The innovation production function remains a bad analytical tool to shed light on the inside of the black box - further surveys are required to deal with knowledge management practices.

3. THE ANALYSIS OF KNOWLEDGE MANAGEMENT: THE PRACTICES VIEW

The literature on capabilities or competencies deals with how some firms can acquire and retain a sustainable competitive advantage (cf. Foss, 1997 for a survey; De Meyer et al.'s survey, 1999 also includes the sociological point of view). The competencies are considered as specific, that is, hard to transfer and copy to another party and thus are not bought and sold on the market. The input-output accounting is thus pointless unless the use of the different resources is understood. A new kind of survey measuring competencies directly or routines is now available (see Appendix 1, Table 7 for France). The Competencies questionnaire (see Appendix 1, Table 6 for general information) is oriented toward a more restricted field of the competencies needed to manage innovation; that is new knowledge. This section begins with a quick presentation of the design and topics of the Competencies questionnaire. The questionnaire is linked then to the existing theoretical and empirical literature on competencies for innovation. The third part presents the main results that are available and limitations are presented in a final sub-section.

3.1. The Topics of the "Competencies" Questionnaire

The Competencies survey aggregates 73 basic competencies into the following 9 categories (see Appendix 2): the capacity to insert innovation in the overall business strategy (set C1); anticipating and adapting to market changes (C2); the development of innovation (C3); the production of innovation (C4); the absorption of external knowledge (C5); the capacities of appropriation (C6); the Human Resources Management (C7); the funding for innovation (C8); and the selling of innovation (C9). The questionnaire's division into nine categories results from a multidisciplinary study involving economists, managers, sociologists and a set of interviews testing the different steps of the questionnaire (see François et al., 1999).

The main focus of the Competencies survey is to identify practices implemented by innovative firms at the internal organizational level[11]. It gathers the different case studies and more recent econometric studies where the innovative firm is represented as the pool of a communication system (e.g.

[11]. However, a growing literature is looking at the link between the productivity effects of the Human Resource Management (HRM) practices. The literature is beyond the scope of this chapter which is focused on innovation. However for people interested in measurement and econometric studies on organizational practices see Huselid (1995), Ichniowski, Shaw & Prennushi (1997), Becker & Huselid (1998) and Mendelson & Pillai (1999). On a large sample, Larsen & Foss (2000) relate HRM practices to innovation productivity. However, the new practices are not especially dedicated to innovation.

Kusunoki et al., 1998), an integration process (see Brown and Einsenhardt, 1995, for a survey), an incentive scheme (see Honig-Haftel and Martin, 1993, Cockburn et al., 1999), or even a hierarchical structure or leadership (Iansiti, 1995, Cockburn et al., 1998).

Although the questionnaire does not introduce new questions on R&D partners or innovation partners, several new aspects on absorptive and appropriation capabilities are included. On absorption, the questionnaire includes, as did CIS1 & CIS2, questions on R&D partnerships (e.g. C5L6, C5L7, C5L11 in Appendix 2), R&D contracts (C5L5, C5L12) and makes progress on the support of the assimilated knowledge: human capital (C5L9), immaterial knowledge through license (C5L8), and knowledge embodied in materials (C5L13). Knowledge can be acquired through the marketplace (license, employees, materials) or non-marketplace (some cooperative links). The questionnaire is thus a first attempt to go inside the black box of absorptive capacities of firms. The Competencies survey also goes further than previous surveys when appropriation strategies are considered: legal practices, such as patenting, are seen as a means among several possible means of knowledge retention. In this perspective, the resolution of human "strategic" competencies (question C6L7 or C6L8; C6L9, C6L11, see Appendix 2) is given in detail. Even more original is the question on the monitoring of strategic knowledge (see C6L10) suggesting that firms organize a disclosure capacity as shown by a large amount of work on formal or informal networking (See Von Hippel, 1987). Last, although little space remained for important topics such as marketing or financing of innovation, the foray into the field remains interesting (see C8 & C9 in Appendix 2).

3.2. A Pragmatic View of Competencies

Different kinds of practices are collected by the questionnaire: formal and informal practices, generic or specific as well as different levels and problem solving.

The first response-category column (see Appendix 2) identifies the existence of the competencies at the firm level without any further detail. The revealed capabilities can be embedded in practices, either formal or informal. This highly encompassing design is at once both useful and misleading. For some items, the practice as well as the level of formalization of the practices is included in the question. The three following items C3L1, C3L4, C5L11 (see Appendix 2) are examples of where a type of formal practice is suggested in the answer (teamwork, joint ventures, organization by projects). On the other hand, several questions do not include a degree of formal implementation. A question may be very broad in scope and encompass formal as well as informal routines such as "Do you do R&D in partnership with other firms?" (C5L6).

The second response-category column (see Appendix 2) specifies the capabilities declared in the first response-category column by adding "with the help of specific procedures" to provide a more restricted view of the different practices implemented by firms. The answer discriminates between specific and generic practices. A positive answer provides what a firm has implemented such as a specific practice to manage the specific problem of innovation (see Hölmström, 1989). Therefore the Competencies questionnaire goes beyond the current literature where the distinction between the two kinds of practices is hazy or often considered as specific to the innovation process (e.g. Cockburn et al., 2000).

A pragmatic analytical approach can be added to these suggested different kinds of practices. According to Nonaka & Takeuchi (1995), it is possible to distinguish analytically the level of knowledge flowing between individuals, groups, and organizational capabilities. The survey uses a comprehensive approach. Three levels of knowledge flows can be found among the 72 different items. For example the organizational level is present in question C3L5 ("Do you promote inter-service mobility for innovating?"). Groups of individuals are more likely to be concerned with question C6L11 ("Do you provide incentives specifically for individuals in receipt of your strategic knowledge (remuneration, advancements)?; and finally, individuals are investigated through questions such as C1L7 ("Do you promote the corporate vision with each employee?").

Nonaka and Takeuchi add a fourth level for analysis - inter-organizational knowledge flows and management. The Competencies questionnaire uses a dual strategy to identify the practices dedicated to the external management of knowledge. The first part of the strategy devotes two blocks of questions to the interactions between firms and the knowledge environment in which they operate (C5 and C6 respectively on absorption and appropriation). Furthermore, the questionnaire also explores interactions with financial bodies (see C8L4 in Appendix 2). The second part of the inter-organizational analysis strategy of the Competencies questionnaire falls in the third response-category column (see Appendix 2). This column investigates whether firms have a capability because of persons who do not belong to the firm. The accent placed on the management of external resources is comparable with the CIS1 and CIS2 results in which innovative firms declared that they relied heavily on external sources for innovation. The intersection between the columns and rows is however problematic. An external human resource is obviously often or even always used to cooperate in R&D (see C5L6 in Appendix 2) or when the affiliations of groups are examined (François, 1997).

The initial results, available from this new type of data set, are surveyed then further lessons and limitations are presented.

3.3. Initial Results

The exploitation of the database is scant and scarce. Two different strategies are possible here based on two interpretations of the variables. The first type of investigation considers that a competency is a measurement of performance per se. Then, it is hypothesized, the more numerous capabilities that a firm has, the higher its performance. This type of analysis is applied at the global level (François, 1997) or for the different sub-sets of capabilities (Munier, 1999). A second view (see Lhuillery, 2000) considers that the identified competencies are embedded in organizational practices that are possibly linked to performances.

In the first approach, François (1997) shows that firms are more competent when internal capabilities are considered. (C1 to C4) whereas, difficulties occur when management of the environment is required (absorption, financing, appropriation, marketing i.e. C5, C6, C8, C9 see Appendix 2). At the same time, a general weakness of the communication capabilities with customers or financing bodies is found (Munier, 1999). The author suggests that firms barely master the needs or tastes of customers.

The number of competencies increases with the size of firms. A more original finding is that a non-linearity seems to occur since firms with less than 200 employees declare a number of capabilities lower than do other firms (François, 1997; Munier, 1999). It is also found that SMEs have more difficulties in managing their environment whereas large firms (more than 1000) have more difficulties managing human resources or organizing and conducting knowledge production (C7 & C4). Munier (1999) suggests more precisely that large firms from High-Tech sectors are more able to cooperate or build a joint venture with other firms (or public R&D institutions) as well as to absorb technological knowledge. Large firms especially are more likely to analyze patents or other publications from competitors or even their technology.

Considering competencies as practices implemented by firms, François (1997) gives evidence that two thirds of the competencies are not codified in specific procedures. The rate is 43 percent for very large firms (more than 2000 employees). A striking result here is the high positive correlation between the different practices investigated. The correlation among such an important number of practices is however hard to handle (multi-co-linearity problems occur). Following Ichniowsky et al. (1987), Lhuillery (2000) focuses on rules implemented at the individual level, managing the workers involved in the innovation process (a selection of C3, C4, C5, C6, C8 questions) and investigates the impact on radical product innovation sales. The availability of one competency or in other words, the implementation of one practice often induces a positive effect on innovative performances. Converging with Laursen and Foss (2000), the paper underlines the fact that incentives and training are important HRM dimensions that are often forgotten in the literature. Large firms are more likely to be systematically and "specifically" routinized (in the sense of Column 2 in the questionnaire). Firms belonging to industrial electrical equipment, hardware suppliers, printing, electronic and electric components sectors are significantly more systematically organized. It is also shown that there is complementarity between the different organizational practices implemented, and clustered into coherent organizational systems. Further results suggest that the organizational system of innovation be coherently shaped towards mainly generic practices or mainly specific practices. It is often inefficient to move away from the coherent organizational model. Last, clues are given that a systematic and/or specific organizational system of practices is more innovation productive than a non-systematic and/or generic system and even more productive than a lack of management of the innovation process.

These various initial results point out several limitations of the initiative. This issue is dealt with in a last section.

3.4. Real and Hypothetical Limits

There are several difficulties presented by the statistical studies. Since the weight of each capability is not known, the addition of several competencies (as in François, 1997) is therefore not a reliable measurement of the organizational knowledge of firms. The problem is similar to patent accounting. A Lickert scale could overcome this limitation however it would make questionnaire more difficult to complete.

Many questions led to a very high number of positive responses. How can we interpret such results? Many people would say that the question is too loose or that it cannot be negatively answered by firms (as "do you reward original ideas when they are adopted?" i.e. C4L5). Even if such a bias exists, one can argue that some capabilities are necessary to innovate. Hence, non-discriminating answers can also reflect the essential roots of the innovation process. Of course, such very low variance variables induce econometric problems. The computations of the index on some related questions (as in Greenan and Mairesse on COI, 2000) would be a solution to this problem. A similar problem occurs when the answer is institutionally bounded. For example, in France firms are required by law to re-invest a significant portion of their revenues into training of employees. This would of necessity increase a positive response rate on training.

The Competencies survey is addressed to the innovative firms as well as to the non-innovative firms. The working premise is to identify good practices or competencies implemented by innovative firms. However, the fact that the question is placed first in the questionnaire introduces a selection bias. In addition, non-innovative firms are very likely not to continue once the first question is answered in the negative. A possible solution would be to identify the practices in a first step, and afterward to investigate the outcomes of the innovative process. At this stage, the introduction of non-technological knowledge is useful. However such an extension requires a corresponding effort on the side of possible competencies: marketing, financial or even legal practices would be interesting to include in a further investigation on the management of innovation.

Finally, little is known about the source of the declared competency or practice. When a firm acknowledges a practice there is no method to measure the robustness of the answer. The declared practices might be compulsory and no exceptions allowed. Here, the point of view can change with the degree of centralization of the respondent or his/her position in the hierarchy. For R&D workers that are accustomed to exchanging knowledge with external colleagues, the answer is "yes" whereas the same question is answered in the negative by the chairman... A general solution to this problem would be to address - at their personal address - a questionnaire to different employees chosen at random among the staff of the firm. Even if minimizing response burden is important to statistical offices, the method is an outstanding opportunity to get close to the real innovation and knowledge processes keeping the use of the econometric tools (see Mairesse and Greenan, 1999, on COI survey for such an exercise).

4. CONCLUSION

The nineties have bequeathed to economists and practitioners many results but also many new methodological tools to improve the measurement and the understanding of knowledge production and management at the firm level. Insights seem to be numerous enough to go beyond the usual questionnaires on R&D and innovation and to launch knowledge management practices (KMP) surveys on large random samples.

Pioneer work is without doubt full of pitfalls. Comparing the bulk of surveys and questions investigated to the economic and managerial literature on knowledge, four main caveats need to be mentioned. First, there is no information on the evolution of the codification process and the Competencies survey cannot reveal anything on the trend of codification that are put forward by Foray and Lundvall (1996). As mentioned even dynamic practices can be considered codified knowledge. However, in the Competencies survey, they are measured for one time period and it is not possible to study the adoption or discontinuation by firms of any of the practices over time. A questionnaire on KM practices could introduce two questions to deal with the evolution of practices. A first step is to identify the existence of new practices within a firm. The evolution of practices seems to be paced at a medium term and suggests that an investigation on the previously implemented practices 3 years before the reference year of the survey is reliable and is convenient with other surveys (as CIS). A second interesting step would be to grasp the evolution of existing practices including downsizing, a similar use or a wider diffusion (see the COI questionnaire for such an investigation). Second, no more information is available on the breakdown between tacit and codified knowledge within a firm. Even more, there is no measurement of the codified and tacit knowledge flowing inside and outside the firm through different interaction modes. Such a difficulty is backed by the lack of attention given to the codification process or costs (Levitt and March, 1996) since the codification costs influences the level of codification chosen by firms (David & Foray, 1995). Third, there is also a lack of an indicator on the obsolescence of knowledge in firms. Practices are supposed to keep knowledge and losses of capabilities are not perceived. Such losses go beyond the possible withdrawal of a practice: it is a complementary means to identify inefficient routines. Last, more precise KMP classifications are available in the literature and do not fit to the categories or even vocabulary used at the business level. The questionnaires are thus constrained by the understanding of the different KMP as well as by their diffusion among firms.

The novelty of a questionnaire is not a guarantee of a fruitful treatment of a survey. The scarcity of analysis based on the Competencies survey is one example. The CIS2 or even CIS1 experiences are here also instructive. The lack of data is no longer a good explanation of the lack of results on innovation, even from the international point of view. Two simple reasons are mentioned here: the lack of resources in many statistical offices involved in the management of many surveys; and, often relying on few people to implement a deep statistical treatment of the data sets. The phenomenon is compounded by the lack of academic researchers working on individual data sets compared to the growing number of available variables.

A second important limitation is the availability of data sets. Such a question must be solved ex ante especially when international comparisons are intended. On the last point, a solution can be to rely on micro-aggregation processes as in CIS1 that do not seem to induce too many difficulties (see Mairesse and Mohnen, 2001). Other coordination efforts such as bilateral agreements seem to be burdensome and thus scarcely practiced (see the CIS2 case). If such a minimal level of availability and collaboration is not reached, rough and misleading treatments, results or benchmarking will persist...

Once the different hampering factors have been overcome, the exploitation of the data set would allow us to understand in a first step the different KMP implemented in innovative firms. A second step is of course to overcome the descriptive step of mapping linkages between knowledge inputs, the different outputs of the innovation process and the implemented knowledge management practices. The exercise would allow us to get a richer view of the strategies at the firm level and overcome the usual innovation categories.

APPENDIX 1

Table 6: French surveys on technological innovation at the firm level

Name	Nickname	Managed by	Year	Sectors	Period	Sampling Method	Method	Responses	Response Rate
R&D Survey	R&D	Education Nationale DPD C3	Each Year	All	1973-1999	Exhaustive All size, All sectors	Mail Mandatory	3000-4000	79%[D]
Researchers	Chercheur	Education nationale DPD C3	Each Year	All	1980-1999	Exhaustive All size, All sectors	Mail Non-Mandatory Joined to the R&D survey	3000-4000	59%[D]
Survey on techno. Innovation	EAE 90	SESSI, INSEE, SCEES	1991	Manuf.	1991	Exhaustive >19 employ.	Mail Non-Mandatory	More than 19 000	N[D]
First Community Innovation Survey	CIS1	SESSI, DPDC3	1993	Manuf[A]	1990-1992	Sample >19 employ.	Mail Non-Mandatory	3917[C]	83%[C]
Appropriation Survey	PACE	SESSI	1993	Manuf[A]	1990-1992	Sample >49 employ.	Mail Non-Mandatory	1797	
Second Community Innovation Survey	CIS2	SESSI SCEES INSEE	1997	Agric., Manuf, Some Serv.	1994-1996	Sample >19 employ. in manuf. > 9 employ. in services	Mail Non-Manuf.	3917[C]	85%[C]
Survey on Innovation Compétencies	COMP-ETENCIES	SESSI	1996	Manuf[A]	1994-1996	Sample >19 employ.	Mail Non-Mandatory	4150	83%
Information technology	NIT	SESSI	1999	Manuf[A]	1998	Sample >19 employ.	Mail Non-Mandatory	3400	62%
Financing Innovation	FIC	SESSI	2000	Manuf[A]	1997-1999	Sample >19 employ.	Mail Non-Mandatory	3700	70%

[A] Food industries excepted
[B] Engineering, Telecommunications, Computers, Bank-Insurance.
[C] on the SESSI part only
[D] for 1998

Table 7 : French surveys on organizational innovation or change at the establishment or firm level

Survey	Nickname	Managed By	Year	Sectors	Covered Period	Sampling Method	Method	Number of Valid Responses	Response Rate
Technique & Labor Org.	TOTTO*	INSEE	1987-1993	Manuf.		>49 employees	Interview & Non-Mandatory	1470 ind. 776 ent.	Na.
Professional relation and wage negotiation in enterprise	Response	DARES	1992, 1998	Manuf. &	1990-1992	Sample >49 employees	Non-Mantadory Interviews	2998 employers 3274 employees' representatives	Na.
Organizational change in Production	COP	SESSI	1993	Manuf.	1988-1993	Sample >49 employees	Mail Non-Mandatory	1846 ent.	75%
Organizational Change and Computerization	COI	DARES, SCEES, CEE, SESSI	1997	Manuf.	1994-1997	Sample >19 employees	Mail Non-Mandatory	4283 ent. 8000 employees	88%

APPENDIX 2

Ministère de l'industrie, de la Poste et des Télécommunications
[Department of Industry, Post and Telecommunications]

SURVEY QUESTIONNAIRE FOR
INNOVATION CAPACITY

Ministère de l'industrie
SESSI
Centre d'Enquêtes Statistiques
Rue Claude Bloch
A21 95
14024 Caen Cedex
Fax: 16 (1) 49 19 25 58
Fax: (16) 31457484

Le SESSI
contact for (all information):
Telephone:16 (1) 43 19 45 14
Fax: 16(l) 31457439

PLEASE RESPOND BEFORE

Name of respondent: ....

Title:..

Telephone: ....

Fax: ....

Corporate stamp and authorized signature

This questionnaire is to be completed by SENIOR MANAGEMENT in your company's TECHNOLOGY department. The respondent must know your company's technological activities well and have a general grasp of its employment, investments and business strategy.The information gathered must relate to YOUR COMPANY AS A WHOLE (legal entity), i.e., for all your facilities. If global answers are problematic, please contact the SESSI correspondent (see above) to determine a more suitable data gathering strategy.

Having been accepted by the Conseil National de information Statistique [National Council on Statistical Information], this survey, deemed to be of general interest, is obligatory.Visa n*96 X 066 IN of the Ministère de l'industrie, de la Poste et des Télécommunications and the Ministre chargé de l'Economie [Minister of the Economy] valid for 1996.Pursuant to article 6 of Act n°51711 dated June 7, 1951, as amended, on the obligation, coordination and confidentiality of statistical data, the information gathered through this questionnaire shall in no wise be used for tax audit purposes or economic repression. Also, article 7 of the same Act states that knowingly giving an incorrect answer or failing to answer may be cause for administrative penalty.[Act n°7817 of January 6, 1978, respecting electronic data processing, records and freedoms applies to the answers given in this survey by individual companies. The Act guarantees that respondents have the right to access and correct the information concerning them. This right may be exercised through ...]

Please check all answers that apply

During the three last years (1994, 1995 and 1996), did your company...

		YES	NO
COL1	Make technological innovations to its products?		
COL2	Make technological innovations to its procedures?		
COL3	Make aesthetic and design innovations?		
COL4	Make purely commercial innovations?		
COL5	Become, or start the process for becoming, certified ISO 9000?		

IN THE FOLLOWING AREAS OF COMPETENCE

C1. INNOVATIONS IN THE OVERALL BUSINESS STRATEGY

Assess your company's capacity for change.

				by using specific procedures		by using outside consultants	
		YES	NO	YES	NO	YES	NO
C1L1	Do you control production quality and efficiency?						
C1L2	Do you evaluate potential corporate products technologically?						
C1L3	Do you evaluate potential corporate procedures?						
C1L4	Do you evaluate potential corporate organizational structures?						
C1L5	Do you maintain a technological summary report on the company?						
C1L6	Do you maintain an inventory of employee skills?						
C1L7	Do you promote the corporate vision with each employee?						

C2. MONITORING, ANTICIPATING AND REACTING TO MARKET CHANGE

Monitoring competitive products.

Awareness of latent unsatisfied demand and customer response to innovations.

				by using specific procedures		by using outside consultants	
		YES	NO	YES	NO	YES	NO
C2L1	Do you analyze competitive products?						
C2L2	Do you analyze the patents registered by competitors?						
C2L3	Do you analyze the engineering publications of competitors?						
C2L4	Do you analyze the nature (breakdown) and needs of clients?						
C2L5	Do you gather client reaction through after-sales service or distributors?						
C2L6	Do you deliver your product with a means to gather information on client satisfaction (survey included in the packaging)?						
C2L7	Do you conduct end-user tests?						
C2L8	Do you identify emerging needs or groundbreaking behavior in consumer use?						

C3. DEVELOPING INNOVATIVE CONCEPTS
Influencing the organization and time

				by using specific procedures		by using outside consultants	
		YES	NO	YES	NO	YES	NO
C3L1	Is your company structured around its innovation projects?						
C3L2	Does your company involve all services in a project from its outset?						
C3L3	Do you test the innovative product or procedure in an operational context?						
C3L4	Do you promote teamwork or pooled work for innovating?						
C3L5	Do you promote inter-service mobility for innovating?						
C3L6	Do you analyze the defects and breakdowns of new procedures?						
C3L7	Do you readily acquire new technological equipment?						
C3L8	Do you readily acquire new technological supplies?						

C4. ORGANIZING AND MANAGING THE PRODUCTION OF KNOWLEDGE
Fostering and channelling creativity. Evaluating the results of knowledge production

				by using specific procedures		by using outside consultants	
		YES	NO	YES	NO	YES	NO
C4L1	Do you encourage new ideas?						
C4L2	Do you allow everyone some degree of autonomy for innovating?						
C4L3	Do individual evaluations place value on originality and creativity?						
C4L4	Do you permit creative, but not directly productive, behaviour?						
C4L5	Do you reward original ideas that are used?						
C4L6	Do you give reasons for rejecting ideas?						
C4L7	Do you pool knowledge?						
C4L8	Do you compare your collective production of knowledge with that of your competitors?						
C4L9	Do you evaluate everyone's contribution to knowledge production?						

C5. MAKING USE OF OUTSIDE TECHNOLOGY
Identifying, evaluating and assimilating outside technology

				by using specific procedures		by using outside consultants	
		YES	NO	YES	NO	YES	NO
C5L1	Do you know the technology used by your competitors?						
C5L2	Are you aware of the latest technology (technology watcher)?						
C5L3	Do you test outside technology?						
C5L4	Do you conduct R&D?						
C5L5	Do you subcontract or outsource R&D?						
C5L6	Do you conduct R&D jointly with another company?						
C5L7	Do you conduct R&D jointly with public R&D institutions?						
C5L8	Do you use the inventions of others (patents, licences)?						
C5L9	Do you recruit highly qualified, scientific employees in order to innovate?						
C5L10	Have you purchased parts of companies or entire companies in order to innovate?						
C5L11	Do you participate in joint ventures, strategic alliances or other forms of cooperation in order to innovate?						
C5L12	Are you a subcontractor for high-tech components?						
C5L13	Do you assimilate the knowledge incorporated into innovative equipment and components?						

C6. MANAGING AND DEFENDING INTELLECTUAL PROPERTY
Patents, drawings and models, trademarks

				by using specific procedures		by using outside consultants	
		YES	NO	YES	NO	YES	NO
C6L1	Do you innovate on an ongoing basis and/or are you accelerating your rate of innovation?						
C6L2	Does your choice of whether or not to register your title to industrial property depend on the overall benefit to the company?						
C6L3	Do you integrate the risk of copying and reproduction beginning at the design stage?						
C6L4	Do you keep track of the existence and distribution of copies and reproductions?						
C6L5	Do you use legal action to fight copying and reproduction?						
C6L6	Do you take action to have clients and your distribution network place less value on copies and reproductions?						
C6L7	Do you identify your strategic knowledge and know-how?						
C6L8	Do you identify the individuals in receipt of strategic knowledge?						
C6L9	Do you inform your staff of the strategic and confidential nature of this knowledge?						
C6L10	Do you control the divulgence of strategic knowledge?						
C6L11	Do you provide incentives specifically for individuals in receipt of your strategic knowledge (remuneration, advancements)?						
C6L12	Do you take steps to guarantee the minimal loss of strategic knowledge when an employee leaves?						

C7. MANAGING HUMAN RESOURCES WITH A VIEW TO INNOVATION

Recruiting, evaluating, training

				by using specific procedures		by using outside consultants	
		YES	NO	YES	NO	YES	NO
C7L1	1. Do you scout out the current and future specialists in the market?						
C7L2	2. Does your hiring process include an evaluation of innovation potential?						
C7L3	3. Does your hiring process include an evaluation of ability to work with a team?						
C7L4	4. Do you make individual evaluations and rewarding of the best transparent?						
C7L5	5. Do you make mobility rules transparent?						
C7L6	6. Do you evaluate everyone's training needs?						
C7L7	7. Do you make everyone aware of the need to request and select suitable training?						
C7L9	9. Do you evaluate the spin-offs of training on innovation?						
C7L10	10. Do you reward training that is useful to the company?						

C8. FUNDING FOR INNOVATION

Evaluating the costs and finding financial supporters

				by using specific procedures		by using outside consultants	
		YES	NO	YES	NO	YES	NO
C8L1	Do you think through all the costs of an innovative concept ahead of time?						
C8L2	Do you evaluate the cost of past innovations?						
C8L3	Do you know the private and public means for funding innovation?						
C8L4	Do you provide potential sources of innovation funding with information?						

C9. SELLING THE INNOVATION
An innovative product, the innovative company

				by using specific procedures		by using outside consultants	
		YES	NO	YES	NO	YES	NO
C9L1	Do you have a specific promotional strategy for the new product?						
C9L2	Do you determine the target, media and type of publicity for the new product?						
C9L3	Do you project your company as being "innovative and avant-garde" (offices, communications, publications)?						

PART IV

IMPACTS

Chapter Eight
PRACTICE AND KNOWLEDGE MANAGEMENT

Larry Prusak
IBM Institute for Knowledge Management, Boston

1. INTRODUCTION

In this chapter, I will discuss the word 'practice' in the context of knowledge and knowledge management partially. This will allow us to focus both on tacit and explicit knowledge. But what is a practice? To approach this question, I would like to discuss eight heuristics that carry with them assumptions that are fairly well accepted among knowledge practitioners regarding notions of 'understanding'.

2. APPROACHING PRACTICES

A practice includes what ancient Greek philosophers would have referred to as epistemie - that is, universal rules and codified knowledge. These are certainly part of any practice. It also includes what they would have called techné - the skills we commonly associate with an art or a craft. So, as a beginning example, let's look briefly at a professional group that embodies a practice in the sense of epistemie and techné. Let's take cardiac surgeons. I think it is acceptable to say that such surgeons hold both deeply codified and deeply tacit knowledge.

Now, in any fair-sized American city, if you need a cardiac surgeon, what are the first things that you do? Well, at least in the United States, the first thing you do is check your insurance. Then you ask around to find out who is a good surgeon. Let's now say, as is quite possible, that they all went to the same university for medical training - let's say Harvard University. What is the difference among surgeons, and how can you tell? What is the difference between two surgeons who are both 50 years old, from the same university, and who have read the same textbooks? The difference is their tacit skills. It is in their craft. Some of this you might be able to discern by their reputation, or their success rates with patients. This process brings us to consider what a highly skilled surgeon knows and does.

There is a third word that the ancient Greeks used for knowledge. This was metis and this is what Odysseus knew. What did he know? He was a smart fellow. He was King of Ithaca and had some epistemological knowledge at the time. He certainly had craft because he was a good warrior. He could build ships and so on, but he could do something well beyond that. He had what the late Isaiah Berlin said the great politicians had – metis. Metis is the capacity for the rapid-fire summarization of your environment and to decide what course of action is the best, given the circumstances. Today, what one would think of this as being 'street-smart' or having 'savvy', but it also involves an understanding of goals and consequences.

All of these three notions – epistemie, techné, and metis – are extremely useful in approaching practice as a central element of knowledge management. But there is more. Also key in the development and transmission of craft is legend, stories, and lore. Cardiologists are full of stories. When they speak with one another it is common to hear phrases like "hey, remember that patient" or "when we tried this with that woman who had….". They all know and share stories across hospitals and across schools. Now, is this different from 'a profession'? Yes it is. All the college professors across the country do not form one practice. But the cardiologists do, especially in areas where they are physically co-located.

3. HEURISTICS FOR UNDERSTANDING PRACTICE

So, what do these practitioners know? To approach this question, let's try to summarize using eight heuristics and what they mean. These are ways of understanding knowledge management.

3.1. Groups and Communities

A first heuristic is that knowledge 'clumps' in groups. It is not an individual thing. It is socially constructed and one looks for it in groups of people, in networks, in communities of practice.

As Wharton Professor, Sidney Winter, likes to put it: "where is the knowledge in an organization?" This is a key question in corporate strategy. It is not in a person. It is in the institution – in its groups. I have seen, for example, knowledge communities in huge companies such as Ford Motor Company that can be as large as 400 people. Because of the advent of PCs, people tend to think that, with the aid of a lot of commercial technology vendors, an individual sitting at a computer is a 'knowledge node'. This is not true. Knowledge is socially constructed in groups.

3.2. Connectivity Over Capture

A second heuristic would be in the use of various knowledge technologies that are available. Now it is quite possible to fill large spaces with books, journals, articles, newspapers and so on which all deal with knowledge-based systems and technologies. But connectivity wins over capture. Inter-personal interaction allows – indeed promotes – the development of networks, trust and social capital (and thus, knowledge development, advancement and sharing). These interactions give rise to self-organizing knowledge groups – the way that practitioners (like our cardiologists) and scientists do. But the capture model – which has a long lineage in expert systems and AI (artificial intelligence) is still being pushed by some consulting firms and promotes 'the great repository in the sky' as a way of dealing with complex problems. In this model, the law says 'document everything, put it in gigantic taxonomies and retrieve documents that no one is going to read or learn from'. This is truly a wasted effort. But if you adopt a connectivity model, then people will find each other, talk to each other, persuade each other, and develop norms of activity that actually have value in terms of knowledge sharing.

Let me illustrate with an example.

I have a friend who was given the job of knowledge leader at Johnson & Johnson in 1999 - a very large firm and one man with a small budget. The company, which is large and very decentralized, had 180 profit and loss (P&L) centers. What can one man possibly do with a small budget and 180 P&L centers? Do you sit in your office and read books? He found something useful to do. He found that in each of these 180 groups people were working on their own Y2K problems. They all knew that other people in the firm were working on it too, but they really didn't get together. And no one bothered to bring them together. So he spent his whole budget trying to get these people together. He tried to transform a nascent group into an active group in order share with each other, to talk to each other because the problems they were all addressing were very similar. He thought that simple access equaled value. He was proven incorrect. He initially thought that by putting in the capacity or infrastructure for chatting and data sharing, value and synergies would emerge. But when he did that, nothing happened. Instead, he had to get everyone to talk to each other, to send messages – "… hey New Zealand why don't you talk to Germany?;…Germany why don't you talk to France? … You all have the same issues…" Since he had been with the firm for 20 years he was able to use his considerable social capital. He expended this capital to get them to talk and, slowly, they began to talk to each other without him telling them to do so. In this event, his efforts had a huge payback. The firm's New Zealand operations were about to hire consultants and to buy a package of help that would have cost $U.S. 5 million. Instead they found out that the firm's German operations had already bought it, so Germany shipped it over to them and they saved a large sum. The story got to the Board of Directors -

since people learn through stories - they loved it and they increased his budget in the next year. This story has the advantage of not only being true but also of not being unique.

3.3. Technology or Knowledge?

Another heuristic is if you spend more than one-third of your knowledge budget on technology – which of course Americans love to do – then it becomes a technology project and not a knowledge project. The focus on technology swamps the kind of discussion we are having in this book. It becomes 'what bits and bytes make a good portal?' I know many organizations that say "we know all about this soft stuff, about strategy, culture, economics, and sociology, but we're interested in the portal..." Some say that this is knowledge management. It's not. But as James March at Stanford University puts it, "you need both the poetry and the plumbing".

An issue that is indirectly tied to this is that without a strategy or plan, things can go awry. Practitioners begin to fragment. They take advice from consultants and they get swayed by organizational, political and hierarchical issues.

But a simple and well-stated strategy that can be easily communicated - and around which consensus can be built - can be sketched in a series of questions. Such questions include:

- 'Where is the knowledge you want to work with?' (Is it in a database? Is it in someone's head?);
- 'What do you want to do with it?' (Do you want to leverage it? Do you want to mix it with some other knowledge in order to develop a new product? Do you want to sell it? Do you want to codify it? Do you want to transfer it to a strategic partner?); and
- 'What is the goal?' (efficiency, innovation, commercialization, effectiveness?).

3.4. Leadership and Ambiguity

It is something of a truism that without somebody saying 'this is important' then results won't flow. Thus, distributed leadership is key in the practice of knowledge management. This means that not only is it the C.E.O.'s job to lead, but it is a job that permeates the executive suites, the managerial lunchrooms and the shop floors.

We can strengthen this truism regarding leadership with a notion that the great literary critic – Lionel Trilling – called 'negative capability'. This is a tolerance for ambiguity. This means that if people can broaden their attention beyond immediate quantitative payback, if they are willing to think that there are two or three or four answers to a problem, and if they are willing to say 'I don't know, let's experiment', then an important element of real knowledge management has been broached.

3.5. The Value of Time, Space and Passion

A further useful heuristic concerns value. We regularly talk about transferring people around as a way of transferring knowledge and creating value. In addition, knowledge practitioners now know that there is a real value for time, space and passion.

Without time, you do not develop new knowledge. You can't learn something unless you spend time learning it. This may sound elementary. But it still takes the same amount of time to learn French as it did in the year 1500. You can have a PC, you can have all the technology in the world, but it still takes the same amount of time to learn anything. Herbert Simon talked about how long it takes to really be knowledgeable in a subject. He picked the examples of chess and chemical processes. You can buy CD ROMS about chemical processes and chess, but to learn to play chess and to become proficient it still takes the same time. Of course, there are rare exceptions. When I was 10 years old my neighbor was 11 years old. His name was Bobby Fisher. Bobby did not take 20 years to learn chess - at 11 years old he was beating everyone in New York City. Sometimes God strikes.

Space is another key. Without space - cyber, cognitive, or physical space - nothing happens.

You also need passion. Why is it that some people choose to go to meetings? It is because if you have passion, you want to meet other people who are equally committed. You can communicate better with passion and commitment face to face.

3.6. Proxy Measurements

It is worth highlighting the importance of measuring what practitioners do. Knowledge practitioners regularly come up against people who say 'you have to show us the internal rate of return on a knowledge project'. This is nuts. It can rarely be done directly. But what you can do is put in place a series of proxy measures. For example, every year IBM takes out a newspaper ad in all the major newspapers and says that "we lead the world in technology patents, we are number one". What they are really saying is that 'we know an enormous amount about technology. We can't show you the knowledge we have, but we didn't get so many patents without knowing a lot about technology'. Sony does this with new product introduction. Ford does this with global processes. Drug companies do this with drug approvals and the introduction of new drugs to the marketplace. These are all proxy measures for knowledge and they are important and specific to the context.

Another example is all R&D-intensive firms have Ph.D.s or MDs - this can be a good proxy measure of innovative behavior or potential. And again, another set of knowledge proxies is stories. I think we are hard wired to learn from narratives. We learn best through narrative. Aristotle wrote about this along time ago in the Rules of Rhetoric. The Bible has been a big success at least in terms of sales and influence because it is full of stories of what was then believed to be ethical behavior. The Double Helix, by Watson and Crick, tells us the story of their discovery of rDNA and the ease of their communication has laid the ground for the biotechnology revolution that is now emerging.

3.7. Knowledge Cousins

The heuristic point I want to make is the tremendous value in cousins to knowledge. Knowledge cousins are judgment, creativity, and imagination. There's a huge market for these cousins, which cannot be done solely by computers. You cannot program good judgment, imagination, persuasion, or insight.

3.8. Communication Networks

Finally, I would like to highlight the variables in success among people. Take the Nobel Prize in Science - who wins and who loses? This is interesting. First, you have to eliminate IQ because they are all smart. But the second, more telling, variable is the richness, intensity and density of the communication networks among these people. In America, we think of the lone genius in the garage. None of this is true. These people are not isolated savants - instead they are constantly interacting, arguing, and discussing.

4. CONCLUSION

These heuristics and variables, in my view, are some of the important elements for considering knowledge management in innovative firms, organizations and processes.

Chapter Nine
KNOWLEDGE MANAGEMENT AT NRC:
A Practical Perspective to KM

Kathy Wallace
CISTI, National Research Council of Canada

1. INTRODUCTION

Since its establishment in 1916, the National Research Council of Canada (NRC) has played a number of roles for the Federal Government according to the priorities of the time. Research excellence has always been a primary focus. Until the 1980s, NRC's research generally followed internally generated priorities and was seen as part of the 'public good'. Research teams predominantly worked apart from industry although firms did come to NRC for expert advice and funding through the Industrial Research Assistance Program (IRAP).

When government funding reductions began to hit NRC and other national laboratories in the 1980s, NRC needed to adapt its culture, re-focus priorities and begin to generate more revenue. We began to see research groups working collaboratively with industrial firms and there was also more promotion of NRC infrastructure elements such as wind tunnels and towing tanks. Research programs became more aligned with areas key to Canada's economy.

The current thrust of NRC's programs promotes industrial innovation for greater competitiveness. Building and managing knowledge-based relationships is a significant part of NRC's new role, as is developing synergies in regional and industrial sector communities. The organization is strongly focused on helping Canada develop a knowledge-based economy through science and technology, and recognizes that to realize that goal, NRC's employees and their collective knowledge and experience are key elements. NRC is working towards being more virtually structured as an inter-networked enterprise, a change from the independent labs of the past. A national organization with regional and community presence, NRC works within Canada's innovation system and provides scientific and engineering infrastructure, research and knowledge to public and private sectors. Its strength lies in the breadth of its connections to other innovators and its ability to mobilize those capabilities in response to national and regional goals. NRC also maintains infrastructures on behalf of industry in construction, aerospace, ocean engineering and astrophysics.

NRC's regional presence in Canada has grown over time and increasingly operates from institutes in specific areas of the country chosen for the potential syncrgy created by the proximity to a major research institute. The Council brings a number of assets to these communities and many of the assets fall into what are coming to be known as the 3Cs: concepts (ideas and technologies, driven by innovation), competence (skills and the ability to use them) and connections (strategic relationships, nurtured by collaboration) (Kanter, 1999). NRC's world-class research teams help stimulate ideas and lead in the establishment of partnerships with industry and academia. People with knowledge, creativity and dedication are NRC's strength. Some examples: IRAP, a team of 260 advisors located throughout Canada, many with ten or more years of industrial experience, shares technology-development expertise in local communities and provides cost-shared funding to small and medium sized companies to stimulate innovation. One of the world's largest comprehensive information sources exists in the Canada Institute for Scientific and Technical Information (CISTI) within NRC. CISTI is a document supplier as well as a scientific publisher, and it operates in each of the communities where NRC is present, using a network of highly skilled information specialists.

2. CONTEXT

This chapter will use the phrase 'knowledge management' as it is familiar to many people, although the author believes that while information can be managed, knowledge, especially the kind that is held within people's heads or is part of their experience, cannot. Knowledge sharing, knowledge focus, knowledge creation, knowledge integration or knowledge exploration may be more appropriate terms.

NRC is only at the "beginner" stage of the knowledge management continuum in many ways. Selecting and implementing the best knowledge management approach is a challenge in most organizations. NRC is highly decentralized, large, multidisciplinary, knowledge-intensive and functions in more than one language making the task is even more formidable. Leadership commitment to the process is being formalized at NRC and is in the hands of a Director-General level ad hoc group on knowledge management. The 'good practices' in place have been introduced without dedicated KM resources and in many cases the initiatives have come from the local level rather than through any organization-wide efforts.

The author's perspective is as a knowledge management champion within CISTI, formerly Canada's National Science Library. CISTI is recognized as a comprehensive source of scientific and technical information published in books, journals and reports, but the organization is also a publisher of 14 peer-reviewed research journals and is a leader in technologies that provide clients with timely access to science-related information and expertise. The NRC Archives maintain the records of our organization's history and their collections are also part of CISTI.

In most scientific disciplines new research builds on past research. Records related to past scientific endeavors (books, patents, articles and reports) are key knowledge sources that researchers learn from in order to move towards new areas of innovation. Libraries and librarians have been forerunners in knowledge management, as they have been sharing information and knowledge and helping to codify it for many years now.

3. DRIVERS FOR KNOWLEDGE EXPLORATION

There are many reasons why an organization like NRC needs to focus on knowledge and on intellectual capital. New knowledge and innovation are interconnected and today's work environment, including products, services and e-business, is knowledge-intensive. Sharing knowledge can have the effect of codifying tacit information and making it available to others. Everywhere today there is an overabundance of data and information. Individuals find that there are not enough hours in the day to deal with everything yet their work demands that they make fast decisions. Collaborative technologies can reduce duplication of effort and allow information to be synthesized into knowledge and made more widely accessible. Collaborations / partnerships / alliances / knowledge communities are increasingly associated with success and competitive advantage. KM can also be seen as a strategy for promoting competitive technical intelligence in firms.

Turnover and succession planning are issues for organizations with highly specialized workforces. At NRC, 20-40 percent of the organization's tacit knowledge could disappear as people retire within the next 5 years. Because people tend to leave if they feel that their talent is not being fully used, determining ways to identify, reward and leverage the expertise that the organization holds is important.

Much of the literature related to innovation emphasizes the importance of networks and networking. Networks here refer to the social communication process: the sharing of knowledge, practices, tools, ideas and technologies among communities plus the interaction between individuals to facilitate the diffusion and adoption of new ideas and technologies. Information technology-based networks have their place, but trust amongst individuals is easier to establish when people meet face-to-face. Once trust is established, people tend to share knowledge more willingly.

4. CORPORATE KNOWLEDGE MANAGEMENT THRUSTS AT NRC

A number of new programs or business practices at NRC formalize the organization's commitment to create an excellent work environment that embodies a culture of learning, collaboration, trust and sharing of information. NRC's new Employment Philosophy sets out the organization's intent and describes the culture that it hopes to achieve. Competency-based human resources management is similarly being introduced with the goal of enhancing and sustaining excellence in employees. To support that element of the Employment Philosophy, NRC recently launched LearnNet, an NRC-specific tool for learning and development. LearnNet features detailed information for each competency grouping, a self-assessment questionnaire, a template for individual learning plans, and such learning tools as formal training courses and workshops, relevant literature, Web sites and suggested on-the-job learning opportunities.

NRC's Entrepreneurship Program offers awards to those who share their knowledge and exhibit entrepreneurial values and behaviors. It presents opportunities for job growth through internal assignments, and secondment programs, as well as through Entrepreneurship Leave. With Entrepreneurship Leave, employees interested in creating their own companies based on NRC technology can take leave of up to six months of leave with pay; and /or up to two years of leave without pay to do so. The program offers the safety net of a possible return to NRC should the commercialization initiative not succeed. Through the new business practices, NRC's goal is to foster a knowledge-sharing environment with a focus on continuous learning and development.

4.1. Industrial Secondments

Within a program of industrial secondments, NRC research, technical and management employees work in industry and similar employees from industry come to work at NRC. Secondments to/from industry are not new to NRC, but this initiative formalizes the process. The main purpose of the new program is to promote collaboration with industry and to facilitate knowledge and technology transfer. Secondary objectives are to help employees gain a better appreciation of industry needs, and the industrial research and business environment, as well as industry decision-making around technology commercialization.

4.2. Developing a Practice

NRC has made use of collaborative technologies to increase the efficiency of internal and external knowledge sharing. The organization was an early adopter of e-mail and now virtually all employees have e-mail access. Efforts to standardize on common hardware platforms and software programs have met with some success. Sending and receiving information through attachments is certainly far less complicated and daunting for some than it was five years ago. NRC also supports an integrated technology platform, making use of Microsoft Exchange servers. Some institutes share documents and data through public folders while other groups have adopted commercial software such as Lotus Notes / Domino because of its web interface, greater flexibility and security. Over the next few months IRAP will introduce a new management information system that will have a major effect on information management and knowledge management within their group and, possibly, also on other parts of NRC. The client relationship management software will allow staff to record, access and, in some circumstances, to share knowledge about their clients, with appropriate protection of confidential business information. Because IRAP is a large decentralized group, this system, guided by information-sharing principles and guidelines adopted IRAP-wide, will greatly improve management of client information.

NRC has a strong presence on the Internet, internally and externally. The central intranet site for NRC, called Zone, brings together information from groups that provide services to NRC or have a need to communicate with all staff such as: Finance, Human Resources Branch, Information Management Services Branch, Offices of the two Vice Presidents and CISTI. From the site, links provide NRC employees with easy access to a wide range of knowledge sources including the full text of the Human Resources manual, travel and training-related information, access to internal directories, organization charts, and to CISTI's Virtual Library. The Virtual Library, in turn, provides a portal to CISTI's electronic journals and reference tools plus a range of database and current awareness sources available for end-user searching. It also offers links to non-NRC sites related to various research areas. CISTI is one of several NRC groups that attempts to increase the return on investment from staff attendance at conferences or from work-related trips by sharing reports on their intranet site. The idea is that trip reports contain a wealth of knowledge not written down anywhere else, and that the organization can be more effective and efficient the employees know what each other is doing, what key contacts are being made and what the latest technologies are. E-mail or web-based Newsgroups or discussion groups are in use within several institutes including CISTI and IRAP. These 'virtual communities of practice' provide a forum for questions/answers plus exchange of ideas and sharing of information.

The external web sites describe institute programs, focusing on opportunities for NRC - industry collaborations. The sites often share past successes the groups have had. CISTI's services are all web-based. Clients searching the catalog or the current awareness system, CISTI Source, can place orders online and receive them by fax or through electronic technologies permitted by copyright legislation. The external web sites are vehicles to promote NRC services, programs and expertise throughout the world.

5. EXPERIMENTS IN KNOWLEDGE MANAGEMENT

5.1. Scientists Save the Atlantic Shellfish Industry

Stories are often the best way to illustrate the success or impact of knowledge management (Denning, 2000). Let me give you an example of an early KM success story at NRC. In 1987, hospitals in Atlantic Canada reported a number of related cases of illness and death. The common factor was that the people had all eaten seafood – mussels originating in Prince Edward Island. A team of marine bioscience researchers and a CISTI reference librarian worked cooperatively around the clock as a virtual team, from various geographic locations. The scientists approached the librarian with physical properties of the substance and, after using directories and online databases, the librarian sent back results that matched those criteria. The scientists then reviewed the results, did further testing and came up with additional properties the substance exhibited. At the same time, scientists were sharing information/expertise among themselves, trying to narrow down the possible cause of the problem. The mussel toxin had caused the death of three people and left more than one hundred ill. Through this intense process of knowledge exchange, NRC experts, in collaboration with groups from the Department of Fisheries and Oceans and other organizations were able to determine the cause in only 104 hours: the deadly chemical, domoic acid. Investigative work of this kind normally takes several months to complete. The collaborative effort that brought people with appropriate skill sets, tools and experience together saved time and lives. The mussel toxin story got a great deal of publicity and, thankfully, such situations arise only infrequently, but this story illustrates well how collaboration can make a difference. The work in that particular case is credited with saving the Atlantic shellfish industry and is an early example of good knowledge integration at NRC.

5.2. Technology Clusters

Innovation at the local level based on public and private sector relationships and networks is the essence of technology clusters and NRC has already been a player in Canadian clusters formed in Saskatoon (agricultural biotechnology) and Montreal (biopharmaceuticals). Internationally, organizations in the US, Israel, Taiwan, Korea and France have established community based innovation clusters. Economist Michael Porter was among the first to describe clusters in detail (Porter, 1999). Clusters have a focus, like ocean technologies or forest products, and according to NRC President Dr. A.J. Carty, key components of clusters are:

- R&D, done either within firms or accessed through an organization such as NRC;
- A highly skilled workforce;
- Venture or investment capital from knowledgeable sources;
- A government milieu favorable to growth (tax incentives and regulations);
- Information programs and tools to effect knowledge and technology transfer; and
- Incubators and mentors to nurture new enterprises and provide them with management and marketing skills.

The key to clustering is an efficient linkage among all these building blocks. The technology-driven firms thrive on the synergies they derive from associating together and from their access to resources and information (Carty, 2000).

NRC is now helping several communities in Atlantic Canada to develop clusters focused on e-commerce (New Brunswick), ocean technologies (Newfoundland), life sciences (Halifax, Nova Scotia), wireless technology (Sydney, Nova Scotia) and bioresources and the environment (Prince Edward Island). Still other potential cluster opportunities could take place across Canada if resources become available. NRC's role in clusters is as a magnet, a catalyst and a connector. Because the innovation clusters are both social and economic phenomena, network building is an essential element. Within a cluster, government departments at all levels work in partnership with academic institutions and the private sector. The structure provides the capacity and flexibility for groups to act and innovate rapidly, and the system leverages access to important resources and information.

5.3. Expertise Database

For a number of years before web technology was available, CISTI maintained a database and a batch-produced index known as the Knowledge Source Index (KSI). The database included expertise within NRC, other government departments and some academic institutions in Canada. The KSI activity ceased, at least in part, because the program was labor-intensive and searchability within the index was very limited. In 1998, CISTI began work on a project to create a web-accessible database of expertise available at NRC. The database was released the following year and now has over 1,600 entries reflecting NRC expertise. NRC has slightly fewer than 3,000 total employees located in 17 research institutes and 4 technology centers across Canada. All institutes are now contributing to the database. Participation is voluntary and each institute determines eligibility for submission. Some institutes may decide to include only continuing staff, for example, while others include terms, guest workers and any other expertise they feel is relevant and available. For many institutes the participation rate is in the 80-85 percent range. Because e-mail and web-based forms assist the process, CISTI keeps the currency of entries at close to 90 percent. Approximately 3,000 individuals and organizations from around the world access the Expertise Database each month. The database strengthens communication within NRC by helping researchers and technical officers identify colleagues in other parts of the organization who can contribute to current or future projects. It also helps promote NRC research and expertise externally to potential collaborators and partners. In addition, NRC is a partner within the National Expertise Index on Industry Canada's Strategis website. Entries from NRC's database can also be accessed through the Strategis index, so that someone who finds Strategis and searches the National Expertise Index will also learn of NRC areas of expertise.

5.4. Special Interest Groups

NRC institutes in London, Ontario (Integrated Manufacturing Technologies Institute) and in Montreal (Industrial Materials Institute) have established a total of 10 Special Interest Groups (SIGs). Special Interest Groups bring NRC researchers together with industry counterparts in a model that benefits all members and focuses exclusively upon pre-competitive research associated with the improvement of fundamental manufacturing processes. Each SIG targets a specific branch of manufacturing and participant companies pay membership of between \$10,000 to \$30,000 per year depending on their desired level of involvement. Participation in the SIGs has been excellent and members include raw materials providers and equipment makers, as well as manufacturers and some academics. All have an interest in high quality manufacturing that is as fast and as economical as possible. Participants identify areas of research that they would like the NRC institute involved to pursue. Members meet three times a year to discuss and prioritize the initiatives that NRC staff are pursuing as a result of Group input. Special Projects between NRC and one of the member organizations have frequently been spin-offs of the SIG activities. New knowledge, ideas and capabilities are all shared within the Groups and the Special Projects. The success of NRC's SIG model has also had an impact on employment. Member companies must have Canadian manufacturing facilities and all R&D related to the SIGs is conducted within Canada. Some companies eager to benefit from membership have invested in Canada rather than abroad, creating new Canadian job opportunities.

5.5. Industrial Research Assistance Program (IRAP)

IRAP's goal is to increase the innovation capabilities of Canadian small and medium-sized enterprises (SMEs). Through a network of public and private research and technology-based organizations across the country, IRAP's team of more than 260 Industrial Technology Advisors (ITAs) help industry to access expertise, financial assistance and facilities needed to carry out research and pre-competitive development activities. They share skills and knowledge about all aspects of innovation and also work to establish links with other key players within the innovation process, drawing on the expertise of technology-oriented organizations and NRC institutes. ITAs, who have both business and technical expertise, connect SMEs with the right expertise, resources, technology, sources of capital or potential partners and save SMEs time and money.

5.6. Technology Sector Groups

IRAP has now established 11 Technology Sector groups (Agri-food; Biotechnology; Chemical processing; Communications; Construction; Electronics and Signal Processing; Manufacturing Aerospace; Manufacturing Integrated Technology; Manufacturing and Materials Processing; Software; and Sustainable Development). The sector groups are teams of ITAs and their roles include information sharing, fostering collaboration, provision of connections to Canadian technology sources, professional development and networking. Participants in sector groups are geographically dispersed but are linked through e-mail lists and web-related resources are under development. Each group meets face-to-face at least once annually. Questions from members are posted to the mailing list and are answered by those who can contribute to an answer or want to make a comment.

5.7. The Canadian Technology Network (CTN)

One of IRAP's partners, the Canadian Technology Network (CTN) is another knowledge integrating contact for small and medium sized enterprises. CTN member organizations link federal and provincial labs and agencies, universities, community colleges, industry associations, technology centers and economic development agencies. This network of over 1,000 members provides Canadian companies with access to expertise, advice and information associated with technology and related business challenges. CTN is a people network connected by a shared information system. It is the people who make the difference. The CTN Intranet is a good example of knowledge management. It incorporates Lotus Notes / Domino features that facilitate shared document libraries and shared information on client-related activities. The site also has an electronic discussion board that provides another route for questions/answers among member organizations and advisors.

5.8. Publishing and Indexing

A publisher of Canadian and international scientific research results since 1929 and the foremost scientific publisher in Canada, the NRC Research Press helps researchers to share their knowledge through the process of scientific publication (books, conference proceedings and journals). Through funding from the Depository Services Program of Public Works and Government Services Canada, the 14 peer reviewed journals published by the Press are now distributed electronically without charge to anyone with a Canadian IP (Internet Protocol) address. The full text of journal issues published in the past few years is searchable on the site and current and older publications are also indexed and searchable through traditional abstracting/indexing tools.

The Institute for Research in Construction (IRC), formerly the Division of Building Research, has led NRC in providing clients building codes and standards related to the Canadian construction industry. IRC is the home of the National Building, Fire and Plumbing Codes of Canada as well as model codes dealing with energy conservation and a number of helpful companion Guides. In producing the codes, IRC works collaboratively with industry associations and technology centers. For easier searching and transportability most of the codes are issued in CD-ROM format as well as on paper.

Many of the older series of publications from IRC have been digitized and are available to clients through the Internet. IRC also has created a database of all publications ever authored by researchers within the institute and published in the open literature or issued through NRC. Clickable links in the database lead to the full text if it is available in electronic format. The publications database is updated weekly and is a very efficient means of transferring IRC knowledge and technology to the Canadian construction industry.

5.9. Spinning Off New Firms and Incubating Start-Up Companies

Finding new commercialization strategies for the knowledge and technology developed in NRC labs, and pursuing those strategies to achieve the best possible result for the benefit of Canada, is important to NRC. One avenue for commercialization is the "spin-off", where an employee creates a company in which he/she has a principal financial interest, for the purpose of exploiting technology from NRC. More than 130 firms have been incubated or spun off by NRC, and the new firms employ close to 9,000 people. NRC is attempting to create a culture of entrepreneurship within all its research institutes. Incubator facilities at the Council provide technical support to start-up companies that have spun off from NRC, or that collaborate with NRC researchers. In 1999, an Industry Partnership Facility was opened to allow 15 high-tech companies at a time to occupy space and collaborate with researchers at an Ottawa NRC campus. Incubator facilities exist in Montreal at the Biotechnology Research Institute and are planned for other locations. NRC technologies are also shared through licensing agreements and several hundred licenses for NRC-related technology have now been signed. Others in the field are able to build on work that has already been done, rather than starting from square one.

5.10. Competitive Intelligence

There are several initiatives related to competitive intelligence within NRC. One partnership is in place between IRAP and CISTI and the Canadian Institute for Market Intelligence (CIMI) in British Columbia. A second partnership involving IRAP, CTN, CISTI and the University of Manitoba's Asper Centre for Entrepreneurship has recently been established in Winnipeg. Both are focused on helping ITAs access information and expertise and on providing IRAP SME clients with competitive intelligence assessments / analyses.

NRC has also been involved in collaborative work towards the establishment of an Aluminum Research Centre in Chicoutimi and a Photonics Centre in a location yet to be determined. Both initiatives have included consultations involving government, academic and industrial players in efforts to determine existing knowledge/expertise, levels of interest in the projects and the appropriate 'roadmap' or direction that the initiatives should best take.

5.11. Canada Institute for Scientific and Technical Information (CISTI)

Formerly Canada's National Science Library, CISTI is now one of the world's most comprehensive collections of scientific, technical and medical information. The organization of knowledge and access to it is a key objective for CISTI. With new technologies, the focus is on bringing information to individual desktops, putting the power of information in the hands of those who need it every day. NRC employees have the opportunity and ability to search for themselves. CISTI information specialists are now knowledge partners, leveraging the power of information.

5.12. Institute for Information Technology (IIT)

One of NRC's Ottawa institutes involved in work related to knowledge management related tools, IIT's overall goal is to assist industry through collaborative research and development projects. IIT is one of several NRC institutes that holds a regular series of colloquiums or seminars designed to share information related to current or past projects. The projects that IIT has worked on over the past few years are varied and numerous. Products resulting include: EXTRACTOR, a new software module that automatically summarizes any document by scanning the text to select several short phrases which best describe the topic of discourse; and CLUSTIFIER, which addresses the problem of organizing a collection of documents into groups having similar topics. In supervised mode, a person defines the topics by manually grouping ten percent of the documents and CLUSTIFIER does the rest. In automatic mode, CLUSTIFIER automatically discovers a set of topics around which to group. BIRD is a totally new type of Internet search engine providing a superior solution to the problem searching by topic. A person provides BIRD with one or more web pages on a topic of interest; BIRD then returns a ranked list of other web pages on this topic. BIRD is built on algorithms adapted from bibliometrics, making it an excellent complementary technology for integration into Internet search products. ACORN (Agent-Based Community Oriented Retrieval Network) is a product that can distribute documents to all people in an organization to whom the content is relevant. ACORN solves a problem typical of large, dynamic organizations: the people to whom a particular document is relevant are not easily determined. ACORN is designed for inclusion into intranet products, maintaining profiles of each individual's relevant topics, and updating these profiles based on their responses to documents read and recommendations for forwarding them. These are a few samples, rather than an exhaustive list, of work being done within the Institute of Information Technology and it is quite likely that other institutes at NRC are also working on tool development.

5.13. Goals and Rewards – Expected Results/Benefits

Externally, NRC's knowledge sharing practices are having an impact on the growth of firms and new business ventures. They are also positively affecting local employment, increasing innovation and client satisfaction, reducing re-work and they are accelerating product introduction.

Internally, metrics have not been introduced and there is little feedback available but the hope is that staff morale, compensation and employee retention will be positively affected and that increased productivity and accelerated learning will be evident.

6. CONCLUSION

NRC has a number of good knowledge management-related practices, although at this point, there is no formal strategy for KM in place within the organization. The Employment Philosophy, Competency Project (competency-based human resources management), Entrepreneurship Program and an Industrial Secondment Program are all elements that are helping to build a sharing culture and a work environment with opportunities and support to help staff grow professionally and to reward them. Initiatives such as technology clusters, special interest groups and IRAP's work with small and medium sized enterprises, as well as partnerships between NRC institutes and industry groups, benefit all involved. NRC knowledge is shared and a valuable international network of technical and scientific intelligence helps bring external knowledge into the organization. This knowledge and expertise can also be used to transfer S&T information back to Canadian firms and universities and to leverage new innovation opportunities for Canadian industry internationally. Much of NRC's current focus is on building linkages and working with other players who can make things happen. No one group can do it all on its own. NRC also has found that being known through the success of a partner in innovation is a very positive situation. Tools and various collaborative technologies are helping to enable knowledge sharing.

Chapter Ten
INVESTING KNOWLEDGE IN UNIVERSITIES:
Rethinking the Firm's Role in Knowledge Transfer

David A. Wolfe
PROGRIS, Centre for International Studies,
University of Toronto

and

Matthew Lucas
PROGRIS, Centre for International Studies,
University of Toronto

1. INTRODUCTION

Over the past decade, the industrial economies have experienced a dramatic wave of economic and political change. According to the OECD, most industrial economies are undergoing the transition to knowledge–based economies in which the production, use, and distribution of knowledge and information are critical for the process of innovation and economic growth. Current trends in information technologies and the development of global communications networks accentuate the importance of scientific knowledge for the future growth of the economy. The knowledge–based economy (KBE) highlights the production of knowledge in networked institutional settings, and the ability to distribute that knowledge to the relevant components of the innovation system (OECD, 1996a). In this context, universities occupy a critical and somewhat unique role as the primary source of 'knowledge workers' for the knowledge–based economy, as well as an increasingly central source of the key factor of production in the new economy - knowledge itself. As a survey in *The Economist* pointed out, in this view, the university is portrayed "not just as a creator of knowledge, a trainer of young minds and a transmitter of culture, but also as a major agent of economic growth: the knowledge factory, as it were, at the center of the knowledge economy" (David, 1997, p. 4).

In the knowledge–based economy, firms pursue linkages with other parts of the innovation system that provide them with complementary assets, including universities. These relationships with other firms and research institutions help to generalize the higher costs associated with research and development. Since the mid–1980's large, R&D intensive firms in the US and elsewhere have been reducing the amount of in–house basic research they conduct and relying on universities to provide them with that research. The increased emphasis by government on the importance of knowledge–based innovation, combined with the decline of investment in basic research by private firms, generates greater pressure on institutions of higher education to increase their linkages with private firms. They must achieve a balance between their roles in the development and transmission of knowledge (research and teaching) and their third role, transferring that knowledge to other relevant actors in the innovation system (Branscomb, Kodama and Florida, 1999).

Recent studies in Canada and the US identify a growing range of university–industry transfer mechanisms, including: industry liaison offices (ILOs) in universities, research parks affiliated with universities, university–industry consortia, research institutes and centers of excellence, regional development organizations and spinoff firms. These mechanisms perform a wide range of functions, including the negotiation of industrial research contracts, the identification of opportunities for university research in the marketplace and the facilitation of licensing or patenting of research results or the spinoff of new firms. These developments have been supported by a range of new programs introduced by both the federal and provincial governments to promote the transfer of university–based research and technology to Canadian firms. However, the mere proliferation of these activities should not be equated with an increase in their effectiveness or efficiency (Doutriaux and Barker, 1995, Etzkowitz, 1999).

Consistent with this view of universities as potential 'knowledge factories' for the new economy, many policymakers view universities as largely untapped reservoirs of new knowledge waiting to be taken up by firms and applied. They hope that once this knowledge is harnessed it will fuel innovation within the firm, thereby increasing the firm's productivity and national economic growth. To this end policymakers, business executives, and innovation theorists have devoted considerable attention to the question of how to transfer this knowledge from universities to firms. While the policy incentives promoting university–industry partnerships have increased over the past decade and a half, along with the institutional mechanisms to support those partnerships, our understanding of the process by which knowledge itself is managed in this relationship, i.e., the process by which it is generated and transferred from one partner to the other, has not kept pace.

Part of the difficulty in managing the process of knowledge transfer arises from the fact that much of the knowledge created in universities is either tacit in nature or relies on tacit knowledge to be fully understood and applied. To deploy this knowledge in a commercial setting, firms need to capture both its tacit, as well as its more explicit, or codified, component. The difficulty in doing this arises from the fact that the tacit component of university–based research is embedded in the skills, motivation, goals, equipment, materials, and background knowledge that make up the context of the research laboratory. To understand and appropriate the tacit dimension, firms need to establish a common ground with university researchers around these factors. This would help link the knowledge created in the university to a firm's existing knowledge base - an important factor in the firm's ability to absorb new knowledge. This chapter explores the potential for doing so by exploring the nature of knowledge flows from the firm to the university. After examining some traditional views on the process of knowledge transfer as well as recent research elsewhere, it reports on a current research project that analyzes the nature of knowledge transfers in university–industry partnerships and the conditions that enable firms to deploy this knowledge more effectively in their innovative activities.

2. KNOWLEDGE TRANSFER IN THE UNIVERSITY–INDUSTRY RELATIONSHIP

Many conventional approaches to the issue of knowledge flows between universities and industry treat knowledge itself as a neatly packaged commodity and knowledge transfer as a commercial and legal transaction between clearly defined agents. This view ignores the complex nature of knowledge and the linkages and processes that facilitate knowledge flows across institutional boundaries and enable a firm to absorb and employ that knowledge. Three question should be asked when examining knowledge transfer:

(1) What is being transferred?
(2) Why is it transferred?
(3) How is it transferred?

The first question addresses the nature of knowledge, particularly the forms in which it is embodied and the extent to which it is amenable to codification. The second question focuses on the firm's ability to absorb this knowledge and put it to work. The third question considers the linkages between firms and universities and their roles in facilitating knowledge transfer. Most policymakers, along with a significant number of academics, ignore the first question, take the second as self–evident, and focus solely on the third. However, evidence from a growing number of sources, including innovation studies, policy studies, and the sociology of knowledge, suggest that successful knowledge transfer depends on what type of knowledge is involved and how it is employed. One needs to take account of both the context in which the knowledge is created and the context in which it will be employed. Ignoring the first two questions leads to a gross oversimplification of the process of knowledge transfer, resulting in considerable confusion about where the university fits into the innovation process.

The traditional linear model of innovation describes innovation as a sequential progression through basic and applied research, technological development and finally production (Stokes, 1997: 10-12). Both the early supply push and later demand pull variants make a clear distinction between each step in the innovation process and largely restrict knowledge flows between non–consecutive steps. These steps were often mapped onto an institutional framework and used to argue for a complementary division of labor between universities and industry. Since the mid 1980's many economists and policy analysts have criticized the linear model for promoting an overly simplistic view of the relationship between research, development and commercialization by failing to account for the reciprocal interactions or feedback mechanisms between each of the steps. The model also fails to account for the exchange of tacit, or non–codified knowledge that may be embodied in personnel or equipment which is often required to transform scientific ideas into commercial products.

In response to these criticisms Kline and Rosenberg (1986) proposed a chain–link model of innovation in which each part of the process - research, technological development, marketing, and commercialization - influences and is, in turn, influenced by the others to varying degrees. These influences occur through multiple channels and may be embodied in people, material, equipment, and publications. While supporters of the chain–link model often retain the institutional distinctions of the linear model, the links between these institutions become more complex: a researcher's sphere of influence extends far beyond the primary function he or she may serve in the laboratory. The model implies that knowledge exchange between all participants in the innovation process is crucial for successful commercialization to take place. It also suggests that the indirect consequences of university research, particularly training and informal information exchange, may be just as important for economic growth as the development of formal university–industry partnerships.

Martin et. al. (1996) identify six key benefits of publicly funded research:

(1) Increasing the stock of useful knowledge;
(2) Training skilled graduates;
(3) Creating new scientific instrumentation;
(4) Forming networks and social interaction;
(5) Increasing the capacity for scientific and technological problem-solving; and,
(6) Creating new firms.

Each of these benefits result from the creation of new forms of knowledge within the university or public research laboratory. All but the last of these benefits are only realized when existing firms absorb and subsequently employ this knowledge. How firms do this is not well understood. Early work on the economics of knowledge reduced knowledge transfer to firms taking promising insights out of the academic literature. It viewed knowledge as information that can be easily and freely collected by interested firms. Nelson (1959) and Arrow (1962) justified public funding of university research by arguing that the non–rival and non–exclusive nature of the information distributed through academic publications allowed all interested firms to appropriate this information without having to invest in its creation. Firms had no incentive to invest as their competitors could easily reap the benefits. This perspective treated knowledge as durable, freely available and costless to transfer. Knowledge transfer became a two-step process in which universities first disseminated potentially useful knowledge and then firms put that knowledge to work.

In contrast, authors writing in the tradition of evolutionary economics describe knowledge as dynamic, often unarticulated, and claim that firms must invest substantial resources to capture and employ it. This view shifts attention from the applicability of knowledge to the processes that enable a firm to successfully absorb and apply that knowledge. Pavitt (1991) argues that inherent in the traditional justification for public funding of basic research is the danger of confusing the notion of science as a public good (i.e., codified, published, easily reproducible) with science as a free good (i.e., costless to apply as technology). He builds on Rosenberg's (1990) claim that to assimilate and benefit from external research firms have to develop a considerable capacity for research themselves. Pavitt argues that knowledge transfers are mainly person–embodied and that policies that attempt to direct basic research ignore the considerable indirect benefits across a broad range of scientific fields resulting from training and unplanned discoveries. This introduces the notion of knowledge as capacity for acquiring and applying research results rather than as an end in itself. In this framework knowledge is the ability to put information into productive use, it provides the ability to understand new ideas and discoveries and places them in a context that enables more rapid application. Knowledge is therefore not a free and durable good, but involves a large tacit component of skill and capabilities that is embodied in people, products, and procedures.

Cohen and Levinthal (1989) similarly argue that firms need to build an internal knowledge base and research capacity to effectively capture and deploy knowledge acquired from external sources. It is in part the overlap between the firm's knowledge base and external research that allows the firm to recognize potentially useful outside knowledge and use it to reconfigure and augment its existing knowledge base. It also means that to ensure maximum benefit firms have to be intimately involved in the research project. Firms that have developed an internal research capacity and have carefully chosen to invest in university research that compliments this capacity may still capture a large percentage of the benefits from the public knowledge they supports.

Just what these benefits are is the subject of much speculation. Several recent efforts offer some insights on the subject. The Yale survey queried 650 R&D managers in US firms, representing 130 lines of business.

The goal was to understand inter–industry differences in sources of technological opportunity. It distinguished between two roles that science plays in supporting innovation: one as an expanding pool of theory and problem–solving techniques deployed in industrial R&D, but not necessarily new science; the other as a direct source of new technological possibilities pointing the way towards new solutions to old problems. The primary finding was that the role of science as a broad pool of knowledge is more relevant than the direct research findings of scientific research. Current research is generally reported as less directly relevant to the innovative activities of industry. Industrial R&D managers value the scientific background and training of their R&D staff more highly than the current research activities of university–based researchers (Klevorick et al., 1995). Nelson expands on this point in suggesting that there are sound reasons why inventive activity is inherently limited in research organizations that specialize in the conduct of R&D, such as university research laboratories.

> *To do effective industrial R&D generally requires knowledge about the technology of an industry that is not taught in school. It also often requires a certain amount of close and not preprogrammable interaction between the lab and client firm or firms, and complementary work and investment on their part. . . Thus effective lab work requires not only industry–specific but firm specific knowledge, and sensitivity of the lab to the needs of its client firm (1996: 62).*

An alternative research approach deployed in a study of industry–academic linkages offers a different insight into the nature of the relationship. This study examined the relationship from the perspective of the innovating organization, focusing on its knowledge requirements and trying to develop a better understanding of the knowledge flows from academia to industry. The study relied on interviews, but to ensure reasonable depth, they carried out a series of interviews with a number of researchers and managers in each firm across three science–related industries, biotechnology, engineering ceramics, and parallel computing. The authors probed both the links that exist between firms and universities and the types of knowledge flowing to the firms. They also attempted to determine the degree of formality of these links and relative importance of tacit versus codified knowledge. While their findings differ slightly by industry they conclude that partnering with universities contributes most to firm innovation through an exchange of tacit knowledge and that the channels for communicating this knowledge are often informal. Such informal linkages, they argue, are both a precursor and a successor to formal linkages and many useful exchanges of research materials or access to equipment take place through "non-contractual barter arrangements." Faulkner and Senker argue that the flexibility inherent in such arrangements promotes the goodwill between partners that supports more formal linkages. Their evaluation of the types of knowledge firms find most useful shows that codified and tacit knowledge are often used in tandem. "While publications are generally a vital first source of information . . . personal contacts provide the tacit knowledge which is essential if researchers wish to implement or adapt experiential practices and procedures,"(1995: 183).

This body of work on the process of knowledge transfer thus draws attention to the complexities involved in exchanging knowledge between different institutional and organizational settings. Influenced by Polanyi's concept of tacit knowledge, summarized in the phrase, "we know more than we can tell," some authors argue that it is extremely difficult to articulate and codify much of the knowledge required to innovate (Nonaka et al., 2000). This knowledge is deeply embedded in the social and physical context in which it is created. As Senker points out, scientific culture has tended to minimize the importance of skills and tacit knowledge for the research enterprise, yet "while tacit knowledge can be possessed by itself, explicit knowledge must rely on being tacitly understood and applied. Hence all knowledge is either tacit or rooted in tacit knowledge,"(1995: 426).

Despite the fact, that it has been less investigated, and much less appreciated, than the tacit dimension of firm–based innovation, the context of a university laboratory is equally shaped by the background knowledge and skills of its researchers, as well as their goals, the instruments, materials, other physical infrastructure, laboratory procedures and financial resources that they use. Together these elements involve a major degree of tacit knowledge. Firms interested in accessing this tacit knowledge base have two options. They can invest in converting tacit knowledge to codified knowledge or they can attempt to absorb tacit knowledge directly. Although firms have made considerable progress in codifying tacit knowledge, Johnson and Lundvall (2001) argue that even the best attempts result in the loss of considerable knowledge. To directly absorb tacit knowledge firms have to build common ground that overlaps the context of the university laboratory with that of their own R&D laboratories, design shops, and manufacturing plants. To understand how university–industry partnerships work one needs to investigate how this common ground is created and how it facilitates the reciprocal exchange of material and knowledge, tacitly embodied in people and technologies or codified in publications. This analysis is at least as important as an examination of the legal contracts and formal expectations binding the partners.

3. METHODOLOGY

Much of the literature on university–industry partnerships focuses solely on what types of knowledge are being transferred from universities to firms and how firms employ this knowledge in the innovation of new products and processes. While providing important information about the types of linkages that exist between universities and firms and the types of knowledge that firms find most useful, they say little about how these partnerships actually work. Surveys and interviews, usually with managers and researchers within firms, have provided a wealth of data that describe their sources of innovation but they have not clarified the process through which useful knowledge is effectively transferred. This is because much of this literature fails to take the implications of the chain–link model seriously and shapes their investigations around a one–way flow from the university to the firm. In doing so, they divorce the processes of knowledge absorption, management, and application from the context of knowledge creation.

These authors also ignore the reciprocal flow of knowledge, expertise, material, and equipment from the firm to the university. These investments in the partnership embody much of a firm's existing knowledge base and their reconfiguration in the university laboratory and subsequent reintroduction in the firm may be one of the ways in which firms absorb new knowledge. They may also provide a crucial common ground between the partners that encourages communication. This common ground may be especially important given that the partners work within different reward structures that direct them toward different short and long–term research objectives. A university–industry partnership is a dynamic relationship with multiple goals and outcomes which influence innovation at multiple sites. One implication of the chain–link model is that the absorption, management and application of new knowledge is influenced by the process of its creation. Rather than limit investigations to the firm's perceptions of, and expectations for, such a partnership, one should probe the partnership from both sides and examine the dynamic interaction between the partners. The relationship should be studied from the inside out, so to speak, focusing on communication and knowledge flows between partners in addition to contractual obligations and specified deliverables. While legal contracts may establish the structure through which knowledge flows occur, numerous reports suggest that successful knowledge transfer requires diverse and flexible linkages between the partners.

The study reported here examines the types of knowledge exchanged between the partners, the types of linkages that facilitate this exchange, and the social and physical factors that encourage communication across these linkages. Data was collected through a series of in-depth interviews with university and firm participants in collaborative projects at the University of Toronto. To permit comparison across different coordinating mechanisms these projects fall into one of four types of relationships: 1) informal collaboration arranged directly between the researcher and the firm, e.g., consulting, contract work, joint research without direct government or university mediation or funding, employment of students, advising or working for company spun out of one's research; 2) formal relationships usually between one firm and university research teams mediated through government or university programs, e.g., the Ontario Research and Development Fund, the Premier's Research Excellence Awards, Canada Foundation for Innovation, university technology transfer and commercialization offices; 3) formal relationships between one firm and multiple university research teams mediated through joint university and firm infrastructure, e.g., industry funded research institutes housed within the university; 4) formal relationships between multiple firms and university research teams mediated through semi–autonomous independent government or industry funded research body, e.g., Centres of Excellence, university–industry research consortia, innovation/incubation centers, science and technology parks.[1.] To minimize some of our variables the study is limited to an examination of research projects in the fields of information technology and biotechnology. Fifty such cases have been identified and initial interviews completed with approximately half of them.

Interviews are conducted with two or three participants from each side of a partnership. Previous experience has shown that interviewing only the principal investigator or management level personnel provides a limited picture of the relationship. This relationship is often sustained through informal interaction between firm technicians and research personnel and university post-doctoral fellows and graduate students. A follow–on set of interviews will be conducted for each case in six months to one year to assess the evolution of the relationship. At this stage the study is exploring the key characteristics of university–industry partnerships that promote knowledge transfer. The interviews were semi–structured with questions divided into two categories. The first covered background by probing the participants' history with the university or firm and their past experience with university–industry partnerships and the commercialization of research results. The second examined the participants' involvement with the project and their interaction with their partners. It investigated the extent to which the participants communicated with their partners, their overall perceptions of the relationship, the benefit they receive from it, the benefits they perceived their partners to receive, and their expectations for future developments. The results presented below confirm much of the literature on the benefits of university–industry partnerships and provide new insights into the social and physical factors that encourage communication.

4. RESEARCH RESULTS

Not surprisingly our preliminary data confirms many of the conclusions reached by other authors especially concerning the benefits to firms of partnering with university researchers. Table 1 lists in order of frequency the top five benefits of university–industry partnerships reported by firm and university participants.

[1.] A more complete survey of both federal (Canada) and provincial (Ontario) initiatives to support university–based research can be found in Wolfe 1999.

Table 1: Most Frequently Mentioned Benefits of University–Industry Partnerships.

Benefits to the Firm	Benefits to the University
-Access to technical expertise	-Leveraging of government funds
-Employee hiring and retention	-Access to technical expertise
-Information about the knowledge frontier	-Access to equipment and material
-Connection to a larger research community	-Source of interesting problems
-Signaling research capabilities and credibility	-Information about the knowledge frontier

4.1. Benefits to the Firm

In all of our cases the firms collaborated with university researchers on particular research problems of commercial interest. The firms entered into these partnerships because they believed their university partners had theoretical and technical expertise that could help resolve these problems or point the firm in a new direction. Access to technical expertise was especially important for smaller firms that do not have the in–house research and testing capabilities of larger firms. This expertise was valued both for short–term problem solving and long–term technical decisions. University researchers also saved the firm time and money by steering it away from technically impossible or already failed innovations. One important benefit to a firm resulted from university researchers reorganizing and reinterpreting some of the problems the firm was trying to address. Describing a productive meeting between firm managers and university researchers our interviewee commented:

> *That was one example of a good one because what happened in that project is that the problem domain was completely transferred to the research domain who organized the problem domain so it made sense, played that back to the people who had the problem, who said, 'Oh, okay, now I see my problem.' . . . These people came to see the commercial opportunity and the problem domain much clearer than they could have in any other way. No call for papers would give an intimate look at that kind of problem. . .*

This example was particularly valuable because of the way it changed the mental frame for viewing a set of problems and potential solutions. Access to technical expertise was often the primary reason for partnering and was viewed as the direct benefit of collaboration.

The indirect benefits cited by firms were also numerous and diverse. Firms often used partnerships to signal to prospective employees the firm's commitment to research and to give graduate students and post–doctoral fellows an introduction to the firm's working environment. These students were viewed as valuable potential employees as much for the specific skills and expertise they obtained while working on the collaborative project as for their general training and education. These skills were viewed as an important complement to the codified knowledge the firm received from the partnership. These skills allowed the firm to smoothly incorporate new employees into its internal research organization without the usual training and adjustment lag. The opportunity to observe the capabilities of potential employees prior to hiring was also valued. Information about recent advances in the theoretical and technical frontier also rated quite high. The partnership with university researchers provided a window that enabled the firm to keep abreast of potentially competing technologies and new areas for product development. This was closely tied to the role of connecting the firm to the larger research community as the partnership also provided the firm with a bridge to other researchers in the university department and around the world. Finally, partnering with respected university researchers was seen to lend credibility to a firm's product development and signaled the firm's capabilities to potential investors and strategic partners through publications and conference presentations. When asked about the most important impact of his university partners one firm participant responded:

> *"Essentially market credibility. We associate our products with world experts to endorse them. . . academic contacts are primarily for endorsement - they are 'market makers".*

4.2. Benefits to the University

Not surprisingly the top benefit of partnering mentioned by university participants was the additional source of funding it brought. Recent government policy initiatives, such as the Ontario Research and Development Challenge Fund, provincial and federal Centres of Excellence, and the Canada Foundation for Innovation that require matching funding from private sector partners constitute a strong incentive in this respect. Furthermore, a perceived shortfall in funding from existing granting councils, such as the Natural Sciences and Engineering Research Council was also mentioned as an incentive to pursue non–traditional funding sources. In some cases, the availability of venture capital also provided an alternative to government funding. Access to technical expertise that complemented the researcher's own was cited as an important benefit. In some partnerships this was especially beneficial for graduate students and post–doctoral fellows who used the resources available to them at the firm for short–term problem solving, as a sounding board for new ideas, and for information about research techniques and equipment. This is closely tied to another benefit of partnering identified by university researchers. The ability of firms to test and manufacture new materials in large quantities was especially beneficial to researchers in biotechnology. When asked what benefits he obtained from his collaboration with private partners, one researcher replied:

> *Certainly from one company what is coming back is their knowledge about high through–put screening which we don't have here, or access to libraries or medicinal chemists, which is just appearing now here, but was sorely missing. More access than knowledge. . . What we learn from another one of our partners is the chemistry, because that's what they specialize in, and access to testing facilities which are much more difficult for us to manage. . . That company also grew all the cells that we used early on.*

Coming to understand the firm's theoretical and technical difficulties also provided university researchers with new and interesting perspectives on research problems. Many researchers benefited from learning about the design and development aspects of product innovation most often carried on in the firm. While information about the knowledge frontier is commonly associated with university researchers, a number of university participants in biotechnology reported that their firm partnerships helped to keep them abreast of the most recent developments in their field, which were actually occurring in the private sector. Monthly newsletters published by some firms and distributed to their various university partners helped connect the researchers to the larger community.

4.3. Types of Linkages Between Partners

The linkages between the partners included informal personal relationships and formal government and university mechanisms. Approximately twenty percent of the cases involved informal relations between university researchers and a company spun off from the university by a colleague or student, thirty percent were mediated through government matching grant programs, i.e., NSERC Industrial Chairs, ORDCF, PREA, twenty–five percent were mediated through university–industry research institutes situated within the university, and twenty–five percent were mediated through provincial and national Centres of Excellence. In the cases consisting of informal relations or mediated through government grant programs, the principal participant on the firm side is usually a senior manager, or in the case of small firms, the President or Vice–President. The primary contact on the university side is the principal investigator. These participants were usually present during the initial arrangements that established the project and were always present at regularly scheduled progress meeting. All decisions pertaining to the overall direction of the research, publication, and financial matters were determined by these principles within any guidelines established by general policies of the university and the firm, although on the firm side there is always the potential for senior people to step in.

The cases mediated through university–industry research institutes and through provincial and national Centres of Excellence have management committees composed of firm and university representatives that deal primarily with the overall direction of the research and financial matters. The principal participants in the specific project on both the firm and university sides were not usually members of these committees. They were responsible for proposing their project to the management committees in order to received funding. Within these institutes and centers, business development officers sometimes worked to introduce potential partners and to coordinate proposals to the management committees.

4.4. Communication Between Partners

When measuring communication we examined four variables: 1) the type of communication, i.e., face–to–face, electronic, etc.; 2) the frequency of communication; 3) what is being communicated; and 4) the personnel level within the organizations where the communication takes place, i.e., who is communicating. All of the projects studied had regularly scheduled formal progress meetings between the senior managers from the firm, the principal investigator and their research assistants. The frequency of these meetings, which were always held in person, varied from monthly to semi–annually. On a more frequent basis there was ongoing e–mail and phone contact between the principal investigator and senior research personnel in the firm. The frequency of this interaction varied from twice a week to biweekly. In those cases where the principal investigator was the primary link between the firm and the university he or she tended to be the channel for most of the communication between the university research team and the firm. In cases in which the project was mediated through a research institute or Centre of Excellence communication was more dispersed throughout the research team on both sides of the partnership. Graduate students and post–doctoral fellows in these settings often took the initiative to contact firm researchers, in some cases three or four time a week. The majority of these communications took place through e–mail. Much of this communication concerned technical details on equipment and materials that the firm had shared with their research partners or that was already common between them.

In the cases in which business development officers were responsible for initiating communication between principals and between university researchers and firm employees, they frequently retained an ongoing role in facilitating communication. Some of these communications took the form of open seminars to report on research findings, while others were one–on–one meetings or small group sessions. These were usually information gathering sessions in which each partner explained their interest in the project to the other and looked for potential synergies.

4.5. Factors Facilitating Communication

The factors that facilitated communication between the partners consisted of the strength of the social relations between them, the formal mechanisms for sharing results, and the incentives to access each other's expertise and capabilities. In most cases the principle investigator knew his or her counterpart from the firm long before the project was initiated. In many of these cases the he or she had previously worked at the firm as a student. In the cases without a preexisting relationship, the principals were introduced by a common acquaintance at the university or by a business development officer working for an industrial sponsored research institute. A number of interviewees commented on the importance of having an intermediary mediate the relationship between firms and universities. That person could be a member of the firm or the university as long as he or she understood both the goals and work culture of each partner and was able to translate between them. An intermediary was particularly useful in avoiding misunderstanding between the partners, acting to bridge the differences in work culture and language.

In those cases where a long-standing relationship existed, the principal investigator exhibited a strong sense of trust and loyalty to his or her partner. When explaining his close relationship with a particular company one researcher commented that, "the company has been generous enough so I would want them to think that I'm a good friend and therefore give them the first crack at a new idea that we had". The researcher linked the firm's generosity to its investments of both money and technical expertise in the partnership. By investing it's expertise in the partnership, the firm developed closer relationships with the researcher, his graduate students and post–doctoral fellows. Commenting on this relationship the researcher said,

> *They offer collaboration, and they just do things for us. They make stuff. They, in a very informal sense, kind of co–supervise some of my graduate students. Everyone I collaborate with has a Ph.D. and has gone through this and my grad students have direct contact with them. They don't go through me when they can just say 'how do you do this,' or 'can you send us one of these,' or 'do you think this will work,' or 'is this a good idea.' There are a number of people in the organization who are very helpful and very responsive in that way.*

Firm expertise, testing equipment, and material provided a focal point around which firms and universities build fruitful research partnerships.

4.6. Building Common Ground between the Partners

A frequent assumption in most of the literature is that research partnerships crystallize around a common set of objectives and expectations. Our research challenges this view. One firm participant emphasized the incongruence of firm and university objectives and the difficulties inherent in bridging the two,

> *The goal of these programs is to include good research and to include good business. We've learned . . . that the word include is an important word. When you sit people down in a room they want whatever item they are talking about to be good business and good research. That never happens. It is either good research or it's good business. It is not both.*

While this shows that university and firm partners often pursed different short and long-term goals it does not imply that these goals are incommensurate. The quest for fundamental understanding does not necessarily clash with the search for useful applications as projects cannot be reduced to a single objective. Partners are often interested in different aspects of a given research problem and thus describe the same project quite differently. One intermediary between firm and university partners described this well:

> *This is a joint hybrid scheme. The university liaison officer will go and explain it to the university as a certain beast. I will go and explain it to my boss as a communications opportunity because she's the director of communications. You coalesce around an idea. You know that old story about the blind guys touching the elephant? It's a deliberate deconstruction of an elephant into the pieces that everyone wants to touch separately. What you really hope is that when they come together they don't argue whether it's an elephant or not, they just say yeah, I touched that part, it's good.*

Despite these differences some common ground between partners is necessary for successful knowledge transfer.

Table 2: Key Factors Influencing Knowledge Transfer

- Clear articulation of each partner's objectives and expectations
- Frequent communication of problems and findings
- Close Interaction of personnel
- Preexisting relationship between principals or mediation between partners by a third party
- The firm's prior knowledge and capabilities
- Investment of firm's knowledge in the university

While the objectives of the firm and university partner do not need to be identical, it is important that each is aware of the other's expectations. An understanding of each partner's interests and goals not only facilitates the practical work of providing useful results, it broadens each partner's understanding of the larger technical and theoretical issues surrounding the project. Confusion over this matter very quickly reduces each partner's confidence and trust in the other. Cases that exhibited frequent interaction between research assistants in the university and front line researchers in the firm exhibited fewer misunderstandings between the principles involved and showed a higher tendency among students to consider employment with the firm. This interaction also builds informal ties between the partners that facilitate the sharing of knowledge and capabilities. Cases in which students worked directly in the firm for a period of summer employment or, more rarely, cases in which a firm researcher was seconded to the university, exhibited less misunderstanding around the different work cultures and language used by the partners. Where this did not occur, it was often the personal relationship between the university and firm principals or the mediation by an information officer that helped to bridge the cultural and language gap. Personnel exchange or close working relationships also facilitate the sharing of know–how and build trust between the partners. A firm's prior knowledge and research capabilities determine the extent to which the firm can contribute to the partnership and be of use to the university research team. Prior knowledge also provided a link between the firm and the researcher and enabled the firm to absorb new knowledge

One of the most interesting sources of common ground this study highlights, and one that is often overlooked, is the skill sets, equipment, materials, data, and procedures that the firm invests in the partnership. These embodied forms of knowledge may facilitate communication between partners by providing a common point of reference and mediate knowledge transfer by becoming a mixing ground for the firm's old knowledge base and the new knowledge created through the research partnership. Table 3 lists a number of knowledge investments by the firm that occurred in our cases.

Table 3: Knowledge Investments in the Partnership by the Firm

- Exchanging Personnel
- Sharing Background Knowledge
- Sharing Equipment
- Sharing R&D Techniques
- Sharing Materials
- Sharing Data
- Providing Testing Facilities
- Providing Technical Expertise

The sharing of a firm's codified knowledge, expertise, materials, data, and equipment with its university partner helped to build common ground between the context of the university laboratory and the firm. This linked the knowledge created in the partnership to the firm's existing knowledge base and helped to bridge differences in working culture, language, and perspectives on the project by providing a space and impetus for communication across institutional boundaries. Learning took place when communication reconfigured and augmented the firm's existing knowledge base by mixing it with the new knowledge being created in the university laboratory. While this most often consisted of personnel communication between the partners, it also took place through shared expertise, equipment and material that embodied a certain amount of the firm's existing stock of knowledge. The application of this reconfigured knowledge base to old and new problems was the first step in the product and process innovations.

5. CONCLUSIONS

This study confirms many of the conclusions in the existing literature regarding the importance of the indirect benefits to the firm from partnering with universities and the importance of transferring tacit know–how along with the codified research results. It sheds some light on how that knowledge is transferred and the factors that encourage effective communication between partners. To access tacit knowledge firms need to create common ground with university researchers around some of these factors. This has most often been done by attempting to realign the motivation and goals of university researchers with the objectives and expectations of the firm. The nature of the university, however, may limit the extent of this realignment. As we noted above, an effective alternative is for firms to build common ground around the skills, instruments, background knowledge, and procedures by investing their own knowledge assets in the partnership.

For the firm to invest its knowledge assets in the partnership these assets must first be transferred into the university. According to our interviews this most often occurs through three media.

(1) People (skills, insight, understanding, vision, etc.);
(2) Products (instruments, equipment, materials, data, reports, etc.); or
(3) Procedures (testing, manufacturing, design, communications, etc.)

The knowledge transferred through these mediums has a number of characteristics. It is:

(1) Embodied in people, products, and procedures;
(2) Interconnected within the firm (internal linkages between assets);
(3) Connected with external assets (bridges to external knowledge); and
(4) Composed of tacit and codifiable elements.

The first characteristic concerns the forms in which assets are exchanged between partners and actively employed within the firm. The interconnectedness of knowledge assets concerns both the organizational nature of innovation and the difficulty in establishing boundaries between many assets. The connection of internal assets with external knowledge is derived from the fact that the firm's knowledge assets are continually reconfigured and augmented by the firm's interactions with competitors, collaborators, suppliers, customers, and public R&D. These knowledge assets are thus closely connected, through not identical, with assets held by other institutions and therefore provide multiple heterogeneous links with the outside world. The tacit and codified nature of knowledge assets attest to the fact that while the firm can communicate much of its knowledge with others, there are limitations on the extent of this communication.

A firm's knowledge base consists of the knowledge assets the firm actively employs in innovation and the ways in which the firm organizes them to that end: it consists of assets and their relations to each other. This knowledge base is unique and dynamic. It distinguishes the firm from other institutions at the same time as it provides numerous linkages with the outside world. These linkages also facilitate the evolution of the knowledge base. A firm learns as it reconfigures and augments its knowledge base. A firm can do this in one of three ways:

(1) By reconfiguring the relations between assets;

(2) By reconfiguring the assets themselves; and

(3) By adding new assets into the mix.

Partnerships with universities can influence all three of these cases, but firms have traditionally looked to them only as a source of new assets. Recent theoretical developments and empirical studies suggest that much is being lost by not considering the other two possibilities. Universities can help to reorganize relations between a firm's knowledge assets and they can also reconfigure those assets. Firms can facilitate this process by investing more of its own knowledge assets in university partnerships or by partnering with researchers who are using similar materials, equipment, and instruments. The firm should view the knowledge returns as the rent obtained from knowledge investments and not just financial investments.

For policymakers intent on maximizing the economic return on public investments in university research it may be best to focus more attention on increasing the number and diversity of investments that firms place in university partnerships. Financial investments alone will not ensure a close or successful relationship. Multiple investments in personnel, material, equipment and codified knowledge with help to establish a number of diverse communication channels between the partners. A heterogeneous communication network provides more opportunities for mixing the firms existing knowledge stock with that created in the university laboratory. This increases the firm's capacity for absorbing the knowledge created through university partnerships and enhances its ability to employ that knowledge in the innovation of new products and processes.

Chapter Eleven
THE GRAMMAR OF PRODUCTIVE KNOWLEDGE

Nico Stehr
Kulturwissenschaftliches Institut, Essen

A man educated at the expense of much labor and time to any of those employments which require extraordinary dexterity and skill, may be compared to one of those expensive machines.
Adam Smith ([1776] 1909: Book 1. [1.]

1. INTRODUCTORY REMARKS

What is the marginal utility of Heisenberg's uncertainty principle or Durkheim's anomie theory? What is the marginal utility of a test that can detect BSE in cows that are still alive, or younger than one year old? Very difficult questions indeed.

This chapter, which attempts to reflect on the properties of knowledge that make knowledge productive, has the following structure: After some general remarks about the changing role of knowledge in the economy, and the different attributes that are commonly assigned to knowledge - as well as an excursus on knowledge in highly differentiate social systems - I discuss a number of reasons why it appears to be superfluous even to reflect on knowledge that is productive. In the final section of the chapter I will try to offer some features of knowledge that, while not inherent in knowledge, make it productive in the sense that such knowledge is capable of serving as an efficient model for reality.

[1.] In recent years, ever since economists and scholars from other social science disciplines began to stress the crucial role of knowledge in economic affairs, a re-reading of some of the classic texts of the economic literature yields a growing number of citations that endorse the role of knowledge; e.g. Engels, Marx/Engels Werke Band 1: 509; Marshall, [1890] 1964:115; Fourastié, [1951] 1954:277). However, the very fact that these references to the economic role of knowledge needed to be rediscovered already indicates that Stigler's (1961:213) disparaging observation about the status of knowledge in economic theory a couple of decades ago continues to have a considerable ring of truth: "One should hardly have to tell academicians that information is a valuable resource: knowledge is power. And yet it occupies a slum dwelling in the town of economics" (also Rosenberg, 1985).

2. FORMS OF KNOWLEDGE AND MODERN SOCIETY

Productive processes in industrial society are governed by a number of factors, all of which appear to be on the decline in their relative significance as conditions for the possibility of a changing, and particularly a growing, economy. These include, for example, the dynamics of supply and demand for primary products or raw materials; the dependence of employment on production; the nature of the manufacturing sector that processes primary products; the role of labor (in the sense of manual labor); the close relation between physical distance and cost and the social organization of work; the role of international trade in manufactured goods; and the nature of the limits to economic growth. More specifically, there is a shift from an economy largely driven by "material" inputs into the productive process to an economy in which transformations in productive and distributive processes are determined by more "symbolic" or knowledge-based inputs and outputs.[2.]

The reason for analyzing the knowledge structure in modern society is the heightened social significance of knowledge in this society, which is the result of the tremendous increase in the overall fund of knowledge. Of course, shared knowledge has always had a function in social life. In fact, one could speak justifiably of an anthropological constant: human action is knowledge-based. This applies, of course, to economic conduct as well (compare, for example, Veblen, [1908] 1919:324-351; Hayek, 1937; Arrow, 1994). All relations among individuals are based on knowledge of each other, as Simmel ([1908] 1992), for example, has so well described.[3.] Similarly, power has

---

[2.] The implications of these transformations are manifold, of course. For example, I have examined the changing relations between economic and ecological imperatives in a knowledge-based economy elsewhere (see Stehr, 2001b).

[3.] A discussion of the "division of knowledge" in the economy - in analogy to the division of labor - can be found in Hayek (1937; see also Helmstädter, 2000). The importance of (subjective) knowledge for economic agents and the coordination of their activities is described by Hayek ([1945] 1948:78-79), for example, in the following terms: "The various ways in which the knowledge on which the people base their plans is communicated to them is the crucial problem for any theory explaining the economic process, and the problem of what is the best way of utilizing knowledge initially dispersed among all the people is at least one of the main problems of economic policy – or of designing an efficient economic system." The economic perspective Friedrich von Hayek has advanced restrictively stresses the role of individual economic actors as knowledgeable agents and the "invisible hand" properties of knowledge. However, some of the main points about the modern, emergent knowledge-based economy are not merely that economic actors are knowledgeable actors, but also that knowledge becomes a crucial productive force, for instance as a basis for self-sustained growth; that knowledge is not only a unique economic commodity as the result of its embeddedness, but also an extraordinary social (collective) fact hardly adequately captured by such dubious specificities as knowledge and information; the issue of the growing importance of knowledge work; and the fact that more or less skillful attempts are made to explicitly manage and protect

frequently been based on advantages in knowledge, not only on physical strength. And, last but not least, societal reproduction is not merely physical reproduction but, in the case of humans, always cultural, i.e. reproduction of knowledge (Elias, 1991).

In science, reflections on knowledge about knowledge have spawned entire disciplines; specialties are devoted to the production and communication of human knowledge about knowledge, the relations between social action and knowledge, the certification and storage of knowledge claims, and the societal distribution of knowledge.

By the same token, the idea that knowledge may be powerful (Bacon, [1620] 1960), tacit (Polanyi, 1967), forbidden (Shattuck, 1996), relevant (Hayek, 1937), objective (Popper, 1973), self-protecting or self-realizing is widely discussed in social theory, and to some extent in economic theory.

But that knowledge is productive - though the term productive is not without its ambiguity - seems to be an almost superfluous, if not meaningless, statement. Knowledge as an accomplishment of human distinctiveness, creativity and ingenuity is by definition productive. In the world of economic conduct, therefore, the benefits of knowledge for productivity (gains), defined for example as the ratio of output to hours worked,[4] are self-evident. Knowledge generally unlocks the secrets of the world, knowledge enlightens, pushes back the boundaries of ignorance and confers authority, if not profit and power.

It is taken for granted that the accomplishments of virtually all forms of human knowledge must necessarily carry with them the attribute "productive." Productivity is essentially inherent in knowledge. It is non-knowledge that is unproductive. As a result, I will have to spend one of the sections of this chapter enumerating reasons why it is unproductive even to reflect explicitly on the productivity of knowledge and the role of knowledge in production.

When it comes to the utility of such knowledge about knowledge, and thereby conceding that knowledge is not homogeneous, it is widely expected and accepted, at least since the age of the Enlightenment,[5] that scientific,

---

the collective knowledge of economic organizations. Precisely these features of the social role of knowledge in the economy and its consequences are not embodied in von Hayek's approach.

[4] The term and measurement (cf. Colm, 1934) of "production" and "productivity" has its own history in economic discourse. In classical economic discourse, for example, in the writings of Adam Smith, Ricardo and John Stuart Mill, the concept of production was largely confined to activities that resulted in the manufacture of useful material objects. Thus, the labor of a shoemaker would be seen as productive and that of a shoeblack as unproductive. That distinction did not remain uncontested: "A shoemaker alters leather, and thread, and wax into a pair of shoes. A shoeblack alters a dirty pair of shoes into a clean pair" (Senior, 1850:51). Thus, productive or unproductive was no longer classified in relation to material objects. Therefore, production becomes defined as the creation of scarce material or immaterial (economic) goods, while the term "productive" generally contrasts/measures input and outputs.

[5] The qualification is necessary as matter of historical record, since social disbelief of "science," its predecessors, and the usefulness of "knowledge" (Erkenntnis) was widespread

technical or expert knowledge as such is superior, more outstanding and, in the final analysis, more practical.[6.] This is an erroneous truism.

Summing up, the productivity, or the lack of productivity, of knowledge is rarely problematized. This may of course be justified, and I will indicate why such disinterest could be justified. At the same time, I will try to show that the lack of curiosity regarding the conditions for the productivity of knowledge signals that unwarranted and unexamined assumptions about the relation between action and knowledge are taken on board.

In a preliminary way, productive knowledge may be defined as capacities for action or as models for reality that have a considerable probability of realization. Productive knowledge is actionable knowledge.[7.] It is knowledge that is socially legitimate as well as of practical, local relevance. Productive knowledge is context-sensitive. And this means that a discussion of the kinds of attributes that convert models for reality into productive or practical knowledge is based on a non-essentialist understanding of the utility of knowledge.[8.]

I begin by referring to the growing understanding of the role of knowledge in modern economic affairs.

---

until two or three centuries ago. By the same token, skepticism toward science, never totally absent, may be on the rise in modern societies.

6. What exactly confers or assures greater utility on scientific and technical knowledge is a contested matter. The range of possibilities extends from institutional guarantees offered by the scientific community to the method used to generate claims and the close conformity of the claim with reality. All cultural criticism that focuses on and warns us about the extraordinary and dangerous powers of modern scientific and technical knowledge - as Martin Heidegger or Max Horkheimer, for example, do - rests on the assumption that modern science and technology are in fact capable of achieving the kinds of nightmares envisioned by their critics or proponents.

7. The brief definition of the kinds of attributes that distinguish productive knowledge already indicate that dichotomies such as "pure" vs. "applied", "original or "fundamental" vs. "practical", "codified" vs. "uncodified", or "curiosity-driven" vs. ""puzzle-solving" knowledge are not, in my view, a key to achieving a better understanding of the attributes of productive knowledge. For example, the commonplace assertion that fundamental knowledge is needed to advance and direct the solving of practical problems (i.e., "Technology means the systematic application of scientific or other organized knowledge to practical tasks" [Galbraith, 1967:12]), I do not consider an invariant proposition that advances inquiry into the conditions of productive knowledge. The association between what counts as basic research and the ability to solve practical problems is, to say the least, complex. Pavitt (1991:112-113), for example, enumerates four patterns of linkages between basic research and technologies.

8. Given the primary task at hand to identify features of knowledge that make such knowledge productive, I refrain from extending my inquiry to an examination and explication of a theory of the "generation of productive knowledge." However, Loasby (1999:130-135) interprets Adam Smith's theory of the division of labor as a theory designed to solve the problem posed by the scarcity of individual knowledge and the advancement of collective knowledge. The development of knowledge can be attributed to the division of labor; thus, the generation of inventions "provides the basis for a comprehensive theory of the generation of productive knowledge" (Loasby, 1999:134). The division or organization of labor is the catalyst of the extension of knowledge. Some of the opportunity costs this in turns generates are found in debates about interdisciplinarity (see Weingart and Stehr, 2000).

3. KNOWLEDGE AND ECONOMIC CONDUCT

It is fashionable today to minimize the importance of the knowledge of the particular circumstances of time and place (Hayek [1945] 1948:81).

If the central premise of many contemporary observations about the growing importance of knowledge as a productive force is accurate, and if one concurrently accepts the observation that the sociology of science in recent years has made considerable strides in our (microsociological) understanding of the fabrication, dissemination and utilization of knowledge, then it is time to couple the micro- and macroeconomic analysis of the productive role of knowledge with our microsociological understanding of the nature of knowledge. Granted, there is still an enormous gap in economics "between the wealth of microeconomic findings, on the one hand, and the understanding that we have of how knowledge is distributed in the economy as a whole and the ways this affects its performance and dynamics, on the other" (Dosi, 1996:86); a beginning has to be made in theoretical efforts to comprehend the role of knowledge in modern economic affairs in general and the nature of productive knowledge in particular.

4. EXCURSUS: KNOWLEDGE AND THE DIFFERENTIATION OF SOCIAL SYSTEMS

One rather mindful examiner of the development of modern societies, Niklas Luhmann ([1992] 1998:88), doubts whether (scientific) knowledge is capable of becoming a factor of (economic) production. Luhmann points to the growing differentiation of modern society into specialized functional subsystems each based on its own logic and media of communication. Although Luhmann recognizes and discusses, in the context of the question of the ability of science to generate predictions valid in other social systems, the ambition of science to "conquer other functional systems," that desire, he maintains, has not succeeded and cannot succeed.

Luhmann is convinced that there are systemic limits to the power of scientific knowledge. The sciences only produce system-specific and relevant knowledge. Such knowledge is universally valid as scientific knowledge, nothing more and nothing less, since it constitutes a specifically conditioned form of knowledge: namely, knowledge conditioned by the science system as (hypothetical) knowledge. Each of the differentiated social systems is responsible for itself. The universality of responsibility for the systemic functions goes hand-in-hand with the specificity of the systemic reference - for example, knowledge in the case of science. Luhmann ([1992] 1998:88) adds, "early socialists recommended considering knowledge as a factor of production, but this was never really accepted in economic theory, because knowledge cannot own things and is therefore unable to take part in value-added-distribution."

Luhmann's objection to the idea that (scientific) knowledge may be seen and analyzed as productive outside of the scientific system - more specifically, as a force of production - is necessarily, as it were, linked to the proposition that the economic system must thereby itself become a scientific system, just as advice offered by science to politics then would amount to scientific decision-making in the political system. Precisely that is what Luhmann means when he speaks of "conquering" another system, and that is not possible.

Within the frame of reference of his own theory of social systems, Luhmann's reservations about the ability of knowledge fabricated in the scientific community to "work" into other social systems may well be adequate and correct. His reservations about the ability of other social systems to affect the code of scientific activity may be equally warranted. But the transfer - or better: the sharing of knowledge generated in science occurs all the time, and efforts to influence the fabrication of scientific knowledge are significant. This is true for past historical periods and may well become an equally forceful process in the future, as problem-oriented research becomes a more typical activity in science.

In the case of the transfer of knowledge from science, the knowledge claims that are communicated and applied elsewhere are not an exact replicate of the knowledge that transgressed the boundaries of science. One cannot deny the dissolution of the boundaries between science and the rest of society. It takes place in many reciprocal ways. Collins ([1985] 1992:165), for example, points to the "continuity of the networks of social relationships within the scientific professions and networks in society as a whole," as well as the "analogy between cultural production in science and all other forms of social and conceptual innovation." Stichweh (1999), on the other hand, refers to the incursion of the boundaries of the economy into the science system, but does not allow for the extension of the boundaries of science into the economy.

Shifting boundaries between science and politics, for example, may also be manifest with respect to the process of the fabrication of knowledge; in particular, the emergence of cognitive closure, consensus formation, or the evolution of uncontested facts in scientific fields may increasingly incorporate non-scientific actors and non-systemic groups. The more or less direct intervention into cognitive processes in science is perhaps most evident in the case of problem-oriented research, such as environmental research, risk and technology assessment.

5. WHY IT IS UNPRODUCTIVE TO REFLECT ON PRODUCTIVE KNOWLEDGE

Knowledge is a peculiar entity. I would therefore like to enumerate some of the reasons that could possibly account for the broad agreement that discourse on the productivity of knowledge is dispensable. The reasons why any explicit discussion about productive knowledge is not really necessary, or is not a top priority on the agenda of social science, are two-fold: I will distinguish between (1) invariant features of human knowledge and (2) reasons that have to do with the ways in which knowledge is seen to operate within the economic system.

Within the first category of reasons, pointing to invariances of knowledge, one finds a number of propositions that all indicate that the very idea of "productive" knowledge is a non-starter. Most importantly, perhaps, and as long as the primary, governing distinction is the one between non-knowledge and knowledge, then knowledge is always more productive, more desirable and more advantageous than non-knowledge. What usually justifies or governs the difference between knowing that we do not know and knowing that we know involves references to ways of validating knowledge claims as knowledge.

Even if the differentiation is not between knowing and not-knowing, but rather between distinctive forms of knowledge, for example, citing the widely accepted ideal typical distinction advanced by Max Scheler ([1926] 1960: 207), namely the differentiation between knowledge of salvation, cultural knowledge and knowledge of domination[9], the common distinction between everyday and specialized knowledge[10], or the difference captured by the dichotomy of natural vs. social science knowledge, these knowledge

---

9. Only the knowledge of domination - the ability to control and produce effects - produced and cultivated for the purpose of 'changing the world' grows, while the knowledge of culture and knowledge of salvation have been successively relegated to the background (cf. Scheler, [1925] 1958:43). Knowledge of domination resonates with the terms of technical and scientific knowledge. The nearly forgotten but once heated debate about the "finalization of scientific knowledge" (that is, that an effective external regulation - instrumentalization - of scientific fields depends on the completion of theoretical work within its boundaries) suggested that external goals can be accommodated more easily and without further theoretical work once theoretical efforts in science have reached maturity and closure (see Böhme, van den Daele and Krohn, 1973; Böhme, 1978). Following the logic of the finalization thesis, one could argue that productive knowledge is a function of the maturity of scientific theories, and would therefore end up repeating the well-known proposition that there is nothing as (privileged in) practical (terms) as a good theory (see Stehr, 1992; Stehr and Grundmann, 2001).

10. The differentiation between layperson and expert, everyday knowledge and scientific knowledge, or life world and scientific community is a rather common, long established asymmetric dichotomy in modern social theory. The premise that expert knowledge is more powerful (if only symbolically), more distant, and incomprehensible to the layperson is taken for granted (e.g. Habermas, 1981:482-488; Beck, 1991:10-18; Bushby, Williams and Rogers, 1997).

categories do not tell us much about their respective productivity in different contexts and why they should differ at all in their respective utility.

As a matter of fact, we can easily imagine, given common preconceptions about the relative efficacy of these different types of knowledge, that all these types of knowledge are productive. What usually governs distinctions of types and forms of knowledge is a reference to their carriers and ultimately to where and how they tend to be generated and certified.[11.]

Furthermore, a discussion of the attributes of productive knowledge could be insignificant, because in reality the utmost priority in social conduct is given to habitual action (Berger and Luckmann, 1967) and not to contemplation, thinking and reflection. Advances in civilization, as Alfred Whitehead, for example, maintains, occur by virtue of extending spheres of habitual social conduct, that is of action that can be performed without prior reasoning. By the same token, Georg Simmel ([1890] 1989:115) refers to what he considers a kind of anthropological constant: "Humans in general are more interested to accomplish something rather than to know how it is done and achieving the former usually preceded insights about the latter" (see also Immanuel Kant's Primat des Willens).

The second set of reasons why it is unproductive to contemplate the productivity of knowledge has to do with the role knowledge typically has within economic discourse. For example, if one emphasizes the extraordinary economic status of knowledge or informational goods, it is argued that all knowledge is - though not exactly to an equal extent, and despite significant investments of resources upfront that lead to its creation - exceptionally productive, since its repeated "reproduction" is inexpensive when compared to the cost of its initial production. Knowledge can be reproduced and sold many times without leaving the control of the original producer/owner.

More specifically, one might ask, what is the marginal productivity of the mapping of the human genome? Thus, some of the additional extraordinary attributes of knowledge, such as the difficulty of fitting knowledge easily into a scheme linking marginal productivity of inputs, relative prices and distributive shares as a precondition for measuring the productivity of inputs, may well constitute a formidable, if not mortal, obstacle to efforts to generate measures of productivity of knowledge inputs.[12.]

The demanding assumptions of the canonical General Equilibrium model in economics refer to additional reasons why it is unproductive to reflect about the role of knowledge, it shortcomings or peculiarities. The

---

11. I exclude here as largely irrelevant to the topic at hand reference to the frequent as well as contested distinction between practical and empirical reasoning. Empirical and practical reasoning are typically distinguished on the basis of their intended purpose: for example, to generate reasons for knowledge and reasons for action (Raz, 1990:15).

12. But such propositions are not necessarily a hindrance to the commonplace call that better knowledge is needed and that more resources should be devoted to research and development.

actors that populate the world of the General Equilibrium model are not only very adept in managing the knowledge and information at hand, but they also have optimal insights into the workings of the economy. The rational ability of individual economic actors to comprehend and master the world in which they live is exceptional; they skillfully exploit opportunities and forecast future conditions. Therefore, as long as economic actors are portrayed as being in manifest control of almost any emerging contingency, the specificities of knowledge, such as its contested nature, its uncertainty and other salient limits, are hardly germane to their world. The invariant attribute of knowledge remains its productivity.

But that the productivity of knowledge is not always guaranteed, and in that sense a permanent, inherent attribute of knowledge, is the subject of the following section.

6. PRODUCTIVE KNOWLEDGE

Given the diversity of knowledge, the range of ways in which knowledge is augmented and acquired, and - last but not least - the diversity of circumstances in which it may be attempted to be realized and/or is seen as legitimate,[13.] it would indeed appear to be impossible to discover invariances that hold across different knowledge forms such as levels of "innovative opportunity" or degree of "cumulativeness" (Dosi, 1996:85), and that make knowledge productive inherently, permanently[14.] and across contexts. Indeed, this would seem to be a demand that can hardly be fulfilled, even if we tend to think of knowledge invariably - for instance, against the background of ignorance - in the highest possible terms.

The skepticism expressed in these observations about the likely absence of intrinsic, permanent features of knowledge that assure its productivity/practicality already indicate that productive knowledge is also, like the very idea of productivity, a relational concept. Thus, if one defines knowledge as a capacity for action or as a model for reality, then productive

---

13. Although I will not explicate the idea in detail, I want to emphasize that productive knowledge claims also resonate with social and cultural peculiarities - not only because productive knowledge may originate among the social and cultural sciences, where issues of legitimacy appear to be a self-evident matter but also in the case knowledge fabricated by the natural sciences. In the case of the natural sciences, the context of augmenting knowledge prior to use for example is both affected by social assumptions, such as the needs of potential users, and subsequently subjected to moral and ethical considerations. In addition, the legitimacy of knowledge includes, as Irwin and Wynne (1996:214), for example, stress, its "accessibility and accountability, whether it implies empowerment or disempowerment of social actors, and its consistency or otherwise with relevant cultural idioms."

14. In a knowledge-based economy, the advantages that may be gained from additional productive knowledge are narrower than in the economy of industrial society, where the cycle that begins with invention and ends in obsolescence took years, at times even decades to complete; this cycle now takes months, weeks or even days.

knowledge refers to knowledge claims that are in fact capable of setting something in motion.[15.]

Since it is impossible to overcome the basic uncertainty associated with any effort to utilize knowledge as a capacity for action, any economic calculation of the productivity of knowledge can only be carried out effectively ex post and not ex ante, that is, only once a context of application has been stabilized, but not before a context stabilizes. As a result, the economic analysis of productive knowledge becomes problematic; but because it becomes problematic, the sociology of knowledge is relevant for generating and augmenting knowledge that may turn out to be productive.

I would like to suggest that reflections about the conditions or constituents of productive knowledge have to start from the assumption that the productivity or adequacy (usefulness) of knowledge, produced in one context but employed in another context, now can be formulated to pertain to the relation between knowledge and the local conditions of action.

Within the context of application, constraints and conditions of action are apprehended as either open or beyond the control of relevant actors. Given such a differentiation of conditions of action, productive knowledge pertains to open conditions of action, which means that theoretical knowledge has to be re-attached to the social context in general, and to those elements of the situation that are actionable in particular.

More precisely, knowledge acquires a productive or active role in the course of social action only under circumstances where such action does not follow purely stereotypical patterns (Max Weber) or is strictly regulated in some other fashion. Knowledge assumes significance under conditions where social action is, for whatever reasons, based on a certain degree of freedom in the courses of action which can be chosen. Karl Mannheim ([1929] 1936:102) defines, in much the same sense, the range of social conduct in which knowledge plays a role as restricted to spheres of social life which have not been routinized and regulated completely. For, as he observes, "conduct, in the sense in which we use it, does not begin until we reach the area where rationalization has not yet penetrated, and where we are forced to make decisions in situations which have as yet not been subjected to regulation." The question of the relation of theory to practice is thus restricted, for Mannheim, precisely to situations that offer a measure of discretion in social conduct and that have not therefore been reduced to a corset of strictly ordered and predictable patterns of social action - although it cannot be ruled out that even under these circumstances, in situations which are repeated with routine regularity, elements of "irrationality" remain.

[15.] The concept of knowledge as a capacity for action signals that the operative difference is not between knowledge and non-knowledge (Wissen and Nichtwissen), but rather between knowledge and action knowledge (Wissen and Handlunsgwissen). Handlungswissen is equivalent to productive knowledge.

7. SUMMARY AND CONCLUDING OBSERVATIONS

Defiantly, knowledge is seen as inherently productive. I have argued, in particular, contrary to essentialist perspectives that the utility of knowledge does not intrinsically reside in knowledge. Knowledge as a capacity for action is neither inherently self-protecting or self-realizing, nor is knowledge immediately productive. Knowledge tends to be localized and its realization embedded in specific technical, institutional, regional and industrial contexts. The productivity of knowledge is specific to industrial sectors, regions and the setting of firms. By the same token, the generation of (new) productive knowledge, in particular, technical knowledge (Antonelli, 1999) is mainly the outcome of localized innovative capacities and learning processes.

Chapter Twelve
KNOWLEDGE, LEARNING AND INNOVATION POLICY[1.]

John de la Mothe
PRIME, Faculty of Administration, University of Ottawa

1. INTRODUCTION

Innovation policy is centrally concerned with stimulating, guiding and monitoring knowledge-based activities within a political jurisdiction, typically a nation or a region. The goals of innovation policy are economic, although they are widely stated in broad welfare terms (e.g. the advancement of knowledge, sustainable development, or social benefits). Its instruments are programs and institutions, as well as ideas. However, as a policy area, not only is it deeply knowledge intensive but its subject itself is knowledge. Hence, given the wide variation of knowledge - from the explicit to the tacit - knowledge management in innovation policy is a domain that is fraught with complexity, ambiguity, uncertainty, judgment, creativity and spontaneity.

Thus meeting the challenges of innovation policy in a dynamic knowledge economy will require substantial advances in our understanding of how our technological, innovation and policy systems interact. This is core to knowledge management. These systems include knowledge producers (such as research groups), knowledge users and appliers (such as firms), knowledge regulators (such as food and drug inspection agencies, intellectual property agencies), knowledge diffusers (including such smart infrastructure as information highways), knowledge funders (such as granting agencies), and so on. Critical, but not central, to these systems are technological trajectories and policy regimes[2.]. Elements of the needed frameworks have begun to emerge over the past decade through the broad view of evolutionary economics and organizational learning. Substantial new attention has also

---

[1.] The research assistance of Tyler Chamberlin and Anastasia Vakos is gratefully acknowledged.

[2.] Here we can differentiate between the scale and scope available to the G-5 (US, Germany, Japan, France and the United Kingdom), as well as between policy orientations such as the dominant but variant social democracies in Europe and Canada which provide a spectrum from tendencies towards more 'American' style laissez-faire political economics to more interventionist social re-distribution type policies.

started to emerge on the front of social capital. Most of that progress, however, has been from a rather high level of analysis - with the possible exception of Robert Putnam's (1994) work on Italian democratic systems. But while much of the recent policy research has usefully begun to look at institutions, processes and practices (Turner 1994), 'equilibrium' states are still an issue. Generally missing are approaches to social dynamics that can complement or embrace emerging theories of endogenous growth and technical change, complexity and networks (Nelson and Winter, 1982; Rycroft and Kash, 1999).

Among those approaches that are dynamic, the most striking common features are learning and knowledge. Learning, as a manifestly dynamic process, provides a powerful counterpoise to theoretical frameworks that privilege power and interest. (These characteristically lead to static prescriptions.) The goal of this chapter is therefore to conceptually inquire whether policy systems - and innovation policy in particular - can learn (cf. e.g. Leeuw, Rist, and Sonnichsen, 2000).

2. CAN POLICY ORGANISATIONS LEARN?

Policy is a process, not a product. Yet in traditional frameworks, the individual policymaker was seen as a rational actor who needed more and better information to make 'better' decisions. The decision maker was portrayed as a thoughtful, lone individual who could be convinced by evidence and who could make sound judgments based on the merits of 'state-of-the-art' knowledge (Vickers 1958). Knowledge therefore directly affected decisions, and decisions were policy. Yet, during the early modern work on innovation policy, there was little attention paid to understanding policy learning processes or the matter in which policy organizations produced, assimilated, used or transferred information (cf. Nelson, 1977).

Over time however, research into innovation policy and knowledge production has grown considerably and has resolutely affirmed that linear models of knowledge production or management are naive at best (Dosi et. al., 1988). The incorporation of knowledge and learning into prescriptions, as well as assessments, of organizational change and performance have reduced the importance of static rational actor models. Besides re-conceptualizing and dynamising the linear notions of knowledge acquisition and learning by policy organizations, institutional theorists have often tended to portray organizations as being deficient at probing, in a substantive way, the root causes of their policy problems. In Canada, for example, questions of productivity, brain drain, avenues to enter the G-5 in terms of R&D spending and performance, the structure of research advice, targeted R&D versus breadth in spending, the benefits of priority setting versus laissez faire, and so on all have a déjà vu quality in that they are on the front policy burner today for innovation policy and knowledge, and have been for more than forty years. Thus innovation policies are sometimes perceived as being limited, as a result of governments simply 'tinkering at the margins' - and as being largely reactive when faced with a problem or crisis.

Such approaches to organizational learning and problem solving (read 'knowledge management') thus ensure the recurrence of similar problems in the future since root causes have neither been identified or squarely addressed. In other words, through such a lens, organizations are not seen as being good learners and therefore, by extension, are seen as being poor knowledge managers. The anecdotal reality is that too many policy organizations - including innovation policy organizations - leave knowledge management and the culture of learning as an ad hoc endeavor instead of an intentional enterprise. This results in a residual build-up – and possible clash - of rules, policies, routines, traditions, cultural and territorial artifacts that affect (or distort) the decision-making process, the interactions between technological-innovation-and policy systems, and ultimately constrain the policymaker. Furthermore, numerous groups within an organization often compete in the production and policy adoption of knowledge agendas, each with their own biases and objectives, and in so doing parochial imperatives can dominate these views and actions to the detriment of policy development.

Thus, as depicted by Chris Argyris (1982), policy organizations have great difficulty in learning and seldom question the underlying basis or the interaction of their own policy problems with other policy groups. Policy organizations have been depicted as lacking in innovative understanding or action and as being resistant to organizational change, implicitly choosing to stress conformity instead of creativity. Indeed, in today's demanding knowledge-based environment, policies and policy organizations are being pressed to be increasingly flexible and responsive. Some have challenged policy organizations to develop more adaptive structures for knowledge management (Thompson, 1969; Bennis, 1971). For policy organizations to learn, they require the requisite skill and incentives to identify and acquire knowledge, to value and store it, to share or transfer it and so on (Dery, 1986). But policy organizations cannot acquire, digest and utilize all the knowledge that is available. As Ryecroft and Kash (2000) point out, organizations must become adept at operating as a networked organization, trusting in tacit knowledge, facilitating interchanges with the external environment and other related institutions, and not isolating policy analysts or actors. Self-inspection has been spouted as one tool to correct organizational deficiencies. But organizational retreats and regular 'public consultations' - which are popular with some government agencies - can sometimes be seen by employees and stakeholders alike with suspicion or cynicism, as a way to reinforce polemic.

Argyris (1982) defines organizational learning in this context as "a process of detecting and correcting error". I would add the need for pro-activity, strategy communication and inclusion of the ranks and would ask at what point in the delivery of a policy (however defined) can 'an error' be either defined or detected. (In other words, benchmarking and evaluation needs to be an integral part of the policy design and development process.) In the day-to-day operation of innovation policy – as policy operatives know - this is far from obvious. But Argyris is right in the sense that policy learning is a process in which an organization continually attempts to become competent in taking action while at the same time reflecting on the action it takes in order to learn. The policy learning process, so enunciated, is thus an iterative process, guided by strategy and vision as well as an awareness of the constituent parts (competing and complementary interests, policies, programs and initiatives), and focused around the management of knowledge which is primed to address and anticipate innovation policy issues.

3. LEARNING IN INNOVATION POLICYMAKING

Of course, organizational learning is a special case of collective learning - one that is of particular relevance to the study of innovation policy, policy change and policymaking. As noted above, classical perspectives on policy defer - not to the privileged place of scientific knowledge (Mulkay, 1979) but - to notions of power, interest and coalitions rather than to learning.

With the rise of formal policy studies and analysis over the past five decades, an excessive emphasis on 'expert' technical knowledge was promoted. Parsons and Clark put it well. In a then-typical view, "political leaders - in creating discrete policy decisions which had well defined alternatives - realized that the increasing scientific and technical content of the decisions they were required to act on required outside advice. They turned to a professional cadre of analysts, who presented the relevant scientific and technical information. The decision-makers, now knowing the relevant consequences of whatever decision they might make, then performed the appropriate balancing of values and interests to arrive at a decision. Note that the analysts operated only in a technical arena and their only audience was the decision maker (1993, p. 16).

Clearly, this caricature is rooted in a hyper-rationalistic version of Operations Research and does not deal well with complexity or ambiguity while it limits itself to problems that have a technical 'fix'. Few real problems fit these parameters, certainly within the realm of innovation policy that - by definition - is complex, knowledge-intensive and ambiguous (Sweeney, 1985). Indeed, if such an apparent separation between the technical and the political were possible, then the tension between Democracy and Reason - of the sort described by John Dewey (1929) - would not arise. But without doubt it is, for most policy analysis, a caricature. In an ambiguous and messy decision environment where policy goals are multiple and contested (i.e., reality), decision makers often do not know what question they need to ask and what the unintended consequences of their decisions may be (Jervis, 1997).

But even with a more realistic view of the relationships between technological, innovation and policy systems, there is still much room for learning in the innovation policy processes. At a macro level, policy analysis and innovation policy frameworks often serve a longer term 'enlightenment function'. 'Innovation Systems' is a case in point. After being widely exposed at Niagara on the Lake in Ontario, Canada in a 1988 conference on Economic Theory and Technical Change (the book with that same title had been edited and published by Giovanni Dosi et. al.), Canadian science and technology policy analysts and senior civil servants actively participated in the OECD's Technology-Economy Program (TEP) (1988-1989) in which innovation systems thinking was propagated. This culminated in Montreal with the final TEP conference which was attended by an array of senior Canadian policy advisors, such as Sylvia Ostry. By 1993, the election statement of the Liberal Party explicitly referred to Innovation Systems as its language and framework for policy, a lexicon that continues today in policy documents and government conferences (cf. de la Mothe, 1995).

But the permeation and adoption of language does not necessarily connote 'learning' in policy. Bureaucracies are quite adept at adopting prevalent policy language while continuing to protect their longstanding programs. An increasing number of studies in policy change have focused on learning as a key - yet under-addressed - issue (cf. Keohane and Nye, 1987). The most common focus is on the learning of top individual policymakers. Steinbruner (1974) describes three characteristic biases in the cognitive processes of organizational decision-making, namely "grooved thinking", "uncommitted thinking", and "theoretical thinking". These, he claims, serve to restrict organizational learning. Others - including Larson (1985), Etheridge (1985), Jervis (1976), Rose (1991), and Ernest Haas (1990) - take a similarly atomistic approach to organizational learning, focusing on the top decision maker and not on the integration and co-ordination of policy streams or the organizational culture which rewards or recognizes learning behavior. Such an orientation, which innovation policy studies are beginning to address in the corporate setting, are therefore promising and will be pursued.

Peter Haas (1990) and Cohen and Prusak (2001) are almost unique in that they take more of a community focus in their work on policy networks in knowledge intensive areas. His 'epistemic communities' are created and dissolved around specific policy issues or crises - such as Mediterranean pollution control - as they arise, but again he gives little illumination as to the learning or adaptive capacity of policy organizations or as to whether policy learning is sustained or temporal.

4. ADAPTIVE LEARNING STRATEGIES OR BEHAVIOURS

The idea of adaptation is not only important in the context of knowledge, learning and innovation policy, but it is also central to evolutionary economics. David Perkins has usefully articulated three different strategies for adaptation that organizations tend to follow in the management of knowledge (Perkins, 2000).

"Adaptation by revision is a basic pattern of search through which humans handle most immediate and accessible needs that require search at all. A longer more descriptive phrase would be 'adaptation by trial and informed revision'", (p. 164). The advantage of adaptation by revision is its directness and efficiency. But Perkins points out that the one limitation of this strategy is its requirement of embedded intelligence. A second limitation is that this behavior can be trapped by 'oases traps' - that is, by local maxima through its persistent efforts to revise and adapt directly based on which it knows. Thus longer-term optimization may not be achieved. This clearly has applicability to policy organizations - such as the Science Council of Canada (de la Mothe, 1992) - which adapt only in a limited non-interactive way.

Unlike the 'adaptation by revision' strategy which has a somewhat Lamarckian quality, 'adaptation by selection' follows a more Darwinian pattern. "From an initial form, variations are generated by a specific mechanism. The variations are constrained within some range - breeding roses does not produce elephants.... But they are not constrained in ways that guarantee improvements in fitness over the initial form... A process of testing ensues, yielding a selection of the more fit variants. The move in Canada towards research partnerships (including matching grants, networks of centers of excellence, and other private-public leveraged funding schemas, might be described as such an adaptation to a changed policy environment).

Adaptation by selection occurs widely... [It] is essentially the mechanism of trial and error or reinforcing learning...

Adaptation by selection has advantages and disadvantages.... On the downside, it is inefficient, investing in many trial variations. On the upside, the process need not embody much intelligence about the fitness landscape.... Thus adaptation by selection can escape.... oases traps", (p. 165).

Adaptation by coding is a strategy that combines the adaptations by revision and by selection. "The general idea is that search involves adaptive forms on two levels that may be called the code and the construction. The code might be a sketch or blueprint, the construction a prototype built from them. The code might be the genome, the construction the organism. The code may determine the construction closely (a blueprint or genetic code) or only loosely constrain it (a sketch or rough idea)...", (p. 166).

To the extent that innovation is seen as not being simply a body of practice but also as a body or culture of understanding - of comprehending - the nature of the learning, evaluating and of the selection processes become increasingly complex. While the notion of 'selection' of 'fit' may speak well in terms of the user and the provider, the realm of policy - and innovation policy in particular - masks numerous levels of impact and interaction.

As Richard Nelson suggests (2000), if the user community is diverse, or there are many different kinds of uses, selection may preserve a wide variety of ideas. On the other hand, the body of understanding that is the result of selection may be quite unified. If both aspects are recognized, it would seem that advances in innovation policy and organizational learning need to be understood as a co-evolutionary process.

5. CO- EVOLUTION, LEARNING AND PRACTICE

Along with learning, co-evolution is the other manifestly dynamic process that amplifies the analysis of change and learning in policy systems. Evolutionary approaches bridge various perspectives to learning in policy outlined above. Evolutionary approaches have begun to be applied to learning processes at the organizational level in innovation policymaking.

At the organizational level, an evolutionary perspective can be fruitful in view of the competitive relations between policy organizations. Nelson and Winter (1982) correctly argue that competition between organizations applies selection pressure to the processes of routine and innovation that operate within institutions.

But in an evolutionary framework, we need a series of learning steps. First we need a description of origination. While many would support the view that particular new ideas are generated by individuals, there is more dispute over the relative effects of individual creativity and social ordering on the total set of new idea or knowledge generation. Second, we need a mechanism of propagation. Different styles of propagation apply to various classes of things learned. Some things need to be replicated cheaply and accurately by print or electronic media. Others, such as behaviors which can be learned by modeling, require more detailed information. Still others, such as skills and craft knowledge, may resist explicit coding and may require direct communication between people, guided participation or practice (Cavali-Sforza and Feldman, 1981). Third, we need mechanisms to introduce variation in propagation. The policymakers' inclination to interpret and modify what they receive in terms of their own experiences, values, information and cognitive processes play an important part here (Fleck, 1979).

But having said this, it is curious that in the organizational learning literature the focus has remained on individuals (including top decision makers, attitude formation, etc.). Group or community theory – including routines and conflict resolution or mediation - is not well developed (and for knowledge management issues to be clarified and advanced community theory needs to be squarely faced). The most frequent approach is not to look at the policy community but instead is to ask how the policy community's learning affects policy change. On this point the theorizing is largely ad hoc (Parson and Clark, 1993). There is, of course, abundant anecdotal evidence concerning the transmission of ideas (Hall, 1989). Here, the adaptation of bodies of ideas is relative. They compete, not against objective criteria but, against each other in view of the opportunities that the environment offers at the moment (Toulmin, 1961; 1972). In innovation policy for example, one can discern a real shift away from linear logics, such as science policy, basic research, applied or strategic research, technology policy, and so on towards more integrative frameworks which require real transparency, co-ordination and functional networks (de la Mothe and Paquet, 1994). But real or authentic operationalization remains truly problematic. It reminds us of the emphasis placed by Rothwell and Zegveld (1984) on the importance of organizational culture and creativity (Nowotny, Scott and Gibbons, 2001) rather than on the adoption of algorithm and design (Schreiber, Akkermans, Anjewwierden, de Hoog, Shadbolt, Van de Velde, and Wielinga, 2000).

If these forces of origination, propagation and variation, coupled with notions of adaptation and selection, imply a strong analogy to biological systems, (Ziman 2001) then we must close off with an idea of the needed counter-balancing effects of stabilization. In any conception of knowledge management, particularly within a policy world, the development of what Fleck called "thought communities" must be considered. Briefly put, stabilization in any organizational context must come from in those intra-personal processes. These include such as discourse, argument, and evaluation and determine what is collectively intelligible and comparative (and therefore what is collectively valuable and manageable). The result is a common or shared body of knowledge. Disciplines provide a stabilization feature in scientific knowledge (Kuhn, 1970). In political, bureaucratic or social discourse, corresponding bodies of thought provide policy characteristics (Geuss, 1981).

But to operationalize these notion of organizational learning with an idea to better fathoming knowledge management in innovation policy, then the question of practice comes into view (Turner, 1994). Like knowledge management itself, the concept of 'practice' is deeply elusive. What are practices? What is being referred to or analytically captured? There are numerous approach to this in order to round out our review. Stephen Turner has usefully illuminated this (p. 2). Philosophically, the immediate review suggests a great deal of overlap: he refers us to Polanyi's 'tacit knowledge', to Ryle's 'knowing how and knowing what', to Kuhn's 'paradigms', to Wittgenstein's 'inherited background', to MacIntyre's and Gadamer's 'tradition', and so on. More sociologically and legalistically thinking - and linking to Weber, Durkheim and Tonnies - we can highly morally binding customs (Sitten)(p. 3) and other related socially binding frameworks. But, from a more evolutionary framework, and in a way that is clearly complementary to these overlapping forays, the notion of embedded routines becomes important for organizational learning, keeping in mind that learning in this specific class of institutions requires sustainably changed behavior.

6. CONCLUSIONS

Knowledge management in innovation policy is thus concerned with the interactions, adaptability and co-evolution between actors and institutions in an attempt to steer, guide and collectively learn within a social economic system. It is centrally concerned with information flows, networks and organization. Such foci of interest hold greater promise than traditional supply-demand rationalizations or top-down management grids that are so popular in political science. Indeed, a learning approach to policy promises to offer, not a solution to contemporary problems but, a framework for effective knowledge management. The emphasis on developing a learning culture, or a knowledge rich culture of practice, is a promising avenue to pursue.

PART V

CONCLUSION

ɔhapter Thirteen
ɔONCLUSION

ohn de la Mothe
RIME. Faculty of Administration
Jniversity of Ottawa

nd

)ominique Foray
ERI, OECD, Paris

As Nico Stehr, in this volume, describes so well, knowledge, for the conomist, has a strange characteristic: sharing or circulating it does not educe the portion everyone has of it but rather allows for multiplying ad nfinitum the number of people who hold this knowledge. The economist xpresses this strange characteristic by qualifying knowledge as being non-ival, meaning that users of knowledge are not rivals. In other words, the use f knowledge by an additional agent does not require producing an additional opy. The marginal cost of using existing knowledge is almost nil (especially n light of the new information and communication technologies). Therefore, ne sharing and circulation of knowledge are actions that make no one poorer, ut on the contrary enrich the communities in which they occur.

Another characteristic of knowledge is that what everyone receives hen taking advantage of the circulation of knowledge is not only a consumer ood (or, to use Machlup's expression, "consumer capital") but also a roducers' good, as the new user can use it to produce new knowledge. It is nis cumulative dimension of knowledge (one idea leads to another, which llows the development of a third) thus making it possible to distinguish etween "superficial" knowledge (small talk and pastime) and scientific or chnological knowledge. Accordingly, the circulation and sharing of nowledge allow not only for increasing the individual enjoyment of it by an nfinite number of agents at almost no cost (when, for example, a poem or a iece of music is circulated), but these actions are also behind the ccumulation of knowledge and the collective progress. They offer countless ew opportunities that are made possible through new combinations, omplementarities, and transpositions of the knowledge being circulated.

ll the contributions in this volume share the point of view that knowledge ust not be equated with information. Whereas information can be transferred at almost no cost (the cost of a photocopy), knowledge must be "reproduced," and this reproduction requires capacities for learning. For example, when you receive a manual on how to operate a machine, you still do not have the knowledge of how to run the machine. But you have a sort of "learning program" that will help you to reproduce the knowledge. There are many situations where the costs of reproducing knowledge are relatively small, particularly when it is circulated among the members of a given community that share a common frame of reference and who have "learned to learn" the type of knowledge that is typically transmitted within the community. In this case, which is a fairly general one, the cost of reproducing knowledge is low, and we are not far from the notion of transfer of information.

Transfer and sharing are therefore essential requirements. As Herbert Simon (1999) put it so clearly, "the evolutionary force of knowledge acquisition depends not only on the speed and cost of creating new knowledge but also on the rate at which knowledge can be transferred."

From this preliminary reasoning we can extract two main ideas.

First, in the knowledge economy, "good fences do not make good neighbors." We have to thank Paul David (2001) for this inversion of the grassroots maxim, "good fences make good neighbors." When two farmers have adjacent fields, one used for growing crops and the other for pasturing livestock, or when gold prospectors are snooping in the neighboring concessions, good fences are a necessity! As David put it, "good fences make probably good neighbors where the resource is land or any other kind of exhaustible resource. But a simple consideration of the public-goods nature of knowledge suggests that this is not so where the resource is knowledge. Knowledge is not like forage - depleted when consumed; data sets are not subject to being "over-grazed" but, instead, are likely to be enriched, rendered more accurate, and more fully documented the more that people are allowed to comb through them."

Put another way, in the case where creating property rights would exclude some potential consumers, the result is wastage. Some will leave dissatisfied, while they could have been so at no cost. Yet the economist detests wastage! Of course, in a dynamic world, knowledge producers respond to incentives. If they cannot enjoy property rights, they will create less, perhaps nothing at all. There is no simple solution to this problem. The answer depends on the situation. What can be said, to close this point, is that the economic issues raised by intellectual property have nothing in common with the issues raised by tangible property.

Second, the transfer and sharing of knowledge are not actions that happen on their own. There must first be rules, standards and incentives of different kinds - in short, culture and organization. For a long time many communities of practices and many communities of professionals incorporated rules and standards for sharing and transfer in their modes of functioning. In other words, they were practicing "knowledge management" without knowing it.

A sharing, exchanging community therefore constitutes the hub or the node in knowledge transmission and circulation systems. Yet these communities are in crisis...and in rebirth, as is evidenced in this volume.

1. IN CRISIS...

If we have never before talked so much about knowledge management, it's clearly because the institutions and behavior that sustain the circulation and sharing of knowledge had shortcomings.

These shortcomings had many, various causes. Dominique Foray (this volume) points out the crisis of internal labor markets and the limitations presented by some professional communities in their capacity to provide a satisfactory means for the circulation and transmission of knowledge. Do we point the finger at inadequate behavior and the lack of a "knowledge-sharing" culture, as mentioned by Meinolf Dierkes (this volume). Without going as far back as the famous weaver from Lyon, De Lasalle in the 18th century, who felt that all that mattered was the circulation of knowledge (he was happy when his innovations were copied), we wonder if some of the basic principles underlying the vertical learning relations between master and apprentice and the horizontal diffusion relations among the members of a given trade have not been seriously shaken up by an onslaught of social transformations (see, for example, Hargreaves, 2000, on the crisis of knowledge circulation in the education sector). In all likelihood, the intensification of competitive attitudes among the members of the same organization and the development of seemingly endless hierarchies and compartmentalization to the nth degree are factors underlying this crisis, as they fragment and dismantle professional communities (Dierkes, this volume).

In the closing address for the conference that gave rise to this book, Keith Pavitt developed another explanation based on one of his favorite topics, namely the incredible degree of specialization of knowledge - an obvious consequence of the Smithian mechanism of furthering the division of labor. While this process can increase economic effectiveness, it also leads to the subdivision and fragmentation of communities. In parallel with this phenomenon, we are also witnessing growth in the number of loci where knowledge is produced. We are witnessing a growing role of users in the production of innovations (von Hippel, 1988), as well as the increased importance of lay people who have become "experts" in their area (Callon, 2001). In short, it is the increased division and dispersion of knowledge that parcels out and fragments communities.

Another phenomenon that fragments and dislocates communities relates back to all the trends to push the envelope of business and piece together complex partnership systems for the sake of improving innovation performance. The modular structure of many products facilitates this type of process (Baldwin and Clark, 1997) by making it unquestionably rational. Modularity can be viewed as being a strategy for organizing an efficient design and production process. Modularity means subdividing tasks and problems into modules. Each module, designed and produced at one location must comply with a number of general rules (architecture, interface), but the module designers are free to experiment with new approaches within these rules. Freedom to experiment is therefore what distinguishes the module supplier from a more traditional subcontractor. We can see, then, that the modular organization is itself an innovation procedure, in that it creates decentralized loci for experimentation while ensuring the integration of innovative modules in the final system.

In addition to the modularization of systems, the outsourcing of functions essential to innovation is an important factor. One need only consider the strategies of the large pharmaceutical firms, whereby they create start-up companies to outsource the creativity function and thus end up in further splitting up of the innovation processes.

In other regards, globalization and the trend to entrust large numbers of transactions to virtual networks are making it increasingly difficult for communities to have a sense of community.

2. IN REBIRTH...

This intense fragmentation of communities, coupled with the decline of some of the institutions that sustain communities, forces the emergence of voluntary efforts to re-assemble communities, to re-weld the links, or even to define the modes for sharing and circulating knowledge while still using fragmented structures.

2.1. Re-assembling the global and local community

Larry Prusak (this volume) develops the following example. In the global firm (J&J in this example), the knowledge manager must strive to assemble a "sense of community," despite enormous spatial constraints (between the centers) and time constraints (which anyone can impose), so that knowledge can circulate and synergies and economies of scale can be exploited.

Re-assembling communities entails an enormous range of possible action. After all, this is the heartbeat of knowledge management. New forms of incentive (rewards, prizes) and new methods of coordination (networks, committees, meetings) are the two pillars of this possible action. The examples of Kathy Wallace (this volume) on the action taken to assemble communities of firms (clusters), as well as the case of SMEs developing network structures, that form the conduits through which information flows (Schuetze, this volume) need to be seen from this perspective.

2.2. Reorganizing the Learning Process

When many scattered sites are involved in a given innovation process (for example a research laboratory, an experimental development site and a prototype production site), it is often simpler and more economical to organize a process of successive iterations among the various sites than to try to transfer and re-assemble all the knowledge at one location. In fact, an important dimension of the knowledge is its non-transferability or at least it may be costly to put this knowledge into a state suitable for transfer. In this case, organizing these successive iterations is a way to create knowledge links among the local communities, with each community contributing to an innovation process (von Hippel, 1994).

Knowledge is not absolute, but rather can only be defined in relation to a specific situation (Tyre and von Hippel, 1997). Such an argument raises the importance of the physical context of learning (Foray, this volume) and, thus, makes it necessary to address the issue of the location of learning within the physical space. In this perspective, Schuetze (this volume) gives the example of product testing in close connection with clients so that better information flows can be developed.

2.3. Linking the Public and the Private

Several contributors (Bordt, Wallace, Wolfe and Lucas, this volume) show that knowledge management is a particularly relevant area of activity in the context of university-industry relations. In this case, the fragmentation of communities becomes all the more intensified as conflicting objectives spring up and each party's incentive structures prove incompatible. It may be noted, however, that the problem is often more one of mutual disinterest in what the other is doing than a strictly economic or ethical one. Academic researchers love to work on "hyper-innovative" solutions but lose interest once the work entails building reliability into the selected solution and overseeing its industrial development. Reliability research is not, in and of itself, a satisfying element for the academic researcher. In comparison, the industrial researcher does not share the same interest for the "elegance" of an innovative solution, as it can be a source of unforeseeable problems and failures. Instead, focus is put on effectiveness in terms of time frames and well-controlled costs. Knowledge management must facilitate the convergence of interests in specific projects in areas where interests diverge spontaneously.

2.4. Circulating Knowledge in the Virtual Worlds

A new type of community has almost come to symbolize the knowledge-based economies, namely virtual communities, whose members meet and interact by using electronic communication methods. Some of these communities are, in and of themselves, knowledge management tools. For example, epistemic communities (Cowan et al., 2000) blend various trades or various services in a single company and work on virtual objects that are, by definition, open and ongoing. This increases everyone's involvement and, in the end, plays a powerful role in the integration and decompartmentalizing of knowledge.

Other communities stand out as examples in the fine-tuning of information exchange mechanisms, such as, for example, the software assistance systems that users offer each other (Lakkhani and von Hippel, 2000).

There are, however, many doubts and questions on how knowledge is circulated and shared in these virtual loci. Does not virtuality give more impetus to the trend toward hyperspecialization by reducing the likelihood of "unscheduled meetings": are not "true libraries" more suitable to unscheduled meetings than virtual libraries (Van Alstyne and Brynjolfsson, 1996)?

3. WHAT CONNECTS KNOWLEDGE MANAGEMENT TO INNOVATION

The link between knowledge management and innovation (or creativity) is evident at a general level, yet becomes more obscure when one attempts to link various types of knowledge management practices to various types of innovation (Landry and Amara, Lhuillery, this volume). An abundance of ideas is a source of creativity. Knowledge management must create this abundance and also organize it.

3.1. Creating Abundance: Exploiting Sources of Knowledge

In the world of mineral and mining resources, creating abundance amounts to discovering new ore deposits and improving the technology so that these new deposits are of economic interest. Abundance is therefore constructed economically and socially (David and Wright, 1991). In parallel, at its core, the creation of knowledge abundance amounts to identifying the new deposits and exploiting them. Yet this is not a simple task.

von Hippel (1988) documents many cases where users are the true innovators. He refers to the field of medical instruments. Typically, users perceive a need to improve an instrument, calibrate the improvement, produce a prototype and disseminate information on the value of the improvement and on how the prototype was made. In this example, the locus of innovation is transferred almost completely to the user level, yet the external link between the user and the instrument manufacturer is essential if the manufacturer is to exploit the user's creative effort and problem-solving.

Yet von Hippel' brilliantly shows that this deposit of knowledge is almost invisible to the firm that supplies the equipment. It is very difficult for the firm to acknowledge that the user (who uses equipment or instruments manufactured by the firm) is a source not only of information but also of innovation. In practice, many trades within the firm producing the equipment, who are in contact with users, have no interest in finding a place in their own organization for the suggestions of users. The employees in charge of maintaining the equipment at the client's site have to carry out standardized maintenance operations that match the standard equipment installed. These employees would therefore be very unhappy to discover that a user has done some tinkering and changed the equipment, which would definitely complicate the maintenance work. They will therefore block any transfer of knowledge. In the same way, the sales manager is interested in obtaining orders for catalogue products from the home plant; there is no interest in looking too closely at the prototypes that users themselves may have developed! In this way, there are many obstacles to these innovations being welcomed by the firm. In this case, knowledge management goes well beyond watching technology: it clearly has to create incentives so that the internal and external sources of knowledge are effectively recognized and exploited. Tracking down emerging ideas and encouraging the members of the organization to express and share their new ideas thus become excellent means to create abundance.

However, creating abundance does not necessarily mean efficiency will follow. Managing the abundance is equally essential. There is a need to go from circulating knowledge to creating an organizational memory.

3.2. Managing Abundance: Creating Organizational Memory

The issues linked to organizational memory and sharing knowledge are particularly important for a firm's innovation performance. Not engaging in deliberate knowledge management strategies may result in significant losses, stemming in particular from redundancies, repetition of errors and shortcomings in the accumulation of knowledge. If problem-solving procedures and creative efforts are conducted exclusively at the local level, there will be some benefit from being in direct contact with the problem to resolve. From another angle, however, keeping problem solving and creativity at the local level increases the risk of producing specific solutions with no input from prior experience, which could potentially have value in addressing the problem. Small organizations, characterized by some job stability, can overcome this problem by developing effective staff networks. But large organizations are confronted by particular difficulties when it comes to re-using existing knowledge to resolve previously encountered problems. There are at least three problems (Steinmueller, 2000):

- First, they must identify the salient features of a given problem to see if it resembles any problems that the organization has seen in the past.
- Second, they must locate the source of relevant information (i.e., the actors who were able to resolve this kind of problem).
- Third, where it is impossible to locate the individual with the requisite knowledge, they must come up with the information through other means.

These three difficulties - the capability of discerning, in a "new" problem, what identifies it with previously encountered problems; the ability to find the individuals who resolved those problems; and in the absence of these individuals, the ability to find the information - are the most common problems of "organizational memory" that large organizations have to confront.

4. CODIFIED AND TACIT KNOWLEDGE MANAGEMENT AND THE ROLE OF INFORMATION COMMUNICATION TECHNOLOGIES (ICTS)

Is knowledge management strictly intended for supporting the formation and circulation of tacit knowledge? We do not think so. Of course, as Prusak explains (this volume), we must probably not technologize knowledge management and spend the equivalent of Portugal's GDP to create "the great repository in the sky"! But we must not mistake the vices of a purely instrumental vision of knowledge management for the virtues of knowledge codification. In fact, codifying knowledge - a process that consists in articulating knowledge such that it can be inscribed on some media - is not just an action that allows for making knowledge storable and transferable at low cost. Codifying knowledge is more a matter of creating new cognitive structures (Goody, 1977). By codifying, we are in fact ensuring "the translation of aural or pictorial expressions into symbolic content. In particular, codification shifts language from the aural to the visual domain, making it possible to arrange and examine knowledge in different ways. Codification provides a spatial device to screen and classify information, opening new opportunities for the modeling or representation of knowledge, a condition for rapid knowledge production and accumulation" (Foray and

Steinmuller, 2001). The codification of knowledge is therefore not just (or even mainly) an action intended to replace tacit knowledge with large knowledge banks. Rather, it is an action of cognitive creativity (a good example would be an engineer, designer or architect scribbling out a drawing and producing knowledge in this way), which constitutes an essential dimension of knowledge management. And from this point of view, it will be necessary to use ICTs to some extent.

5. NO EASY TASK: MANAGING KNOWLEDGE ON A DAY-BY-DAY BASIS!

Some crises or catastrophes, whose solution entails coming up with new knowledge, lend themselves well to effective knowledge management (see the example given by Wallace of the crustacean causing considerable damage to humans, this volume). We also know that in many industries, effective procedures for accumulating knowledge on accidents and sharing it do very well (Cowan and al., 2001).

What is probably more difficult is knowledge management on a day-by-day basis, in peaceful situations, when there are no crises. Everyone thinks that the fact of overlooking a knowledge-management activity will, after all, be of very little importance since it will not have a large impact on the immediate operations of the organization. Therein lies the main danger. While one cannot tolerate disorderliness in the equipment room and while one must not forget to put tools back where they belong, it is very easy to overlook putting one's ideas in order and to forget, for a day or two, to make a record of new knowledge. This is the major difficulty: "cleaning up the ideas room" is a daily activity that costs money and is difficult and unrewarding, for the outcomes are not noticeably spectacular. In fact, no one will notice for a long time to come if the "ideas room" is indescribably messy or properly arranged.

BIBLIOGRAPHY

Abramovitz, M. and David, P.A. (1996). "Technological Change, Intangible Investments and Growth in the Knowledge-Based Economy: The US Historical Experience", in Foray, D. and Lundvall, B.-A. (eds.) *Employment and Growth in the Knowledge-Based Economy*, Paris: OECD.

Acs, Z (ed.) (2000). *Regional Innovation, Knowledge and Global Change*, New York: Pinter.

Acs, Z. and Audretsh, D.B. (1990). *Innovations and Small Firms*, Cambridge, MA: MIT Press.

Adams, J.D. (2000). "Endogenous R&D Spillovers and Industrial Research Productivity", NBER Working Paper, No. 7484, January.

Amable, B. et al, (1997). *Les systèmes d'innovation à l'ère de la globalisation*, Paris: Economica.

Antonelli, C. (1999). "The Evolution of the Industrial Organisation of the Production of Knowledge", *Cambridge Journal of Political Economics*, Vol. 23, pp. 243-260.

Archibugi, D. and Pianta, M. (1996). "Les enquêtes sur l'innovation et les brevets en tant qu'indicateurs de la technologie: état des connaissances", in OECD, *Innovation, brevets et stratégies technologiques*, Paris: OECD.

Archibugi, D. and Sirilli, G. (2000). "The Direct Measurement of Technological Innovation in Business", Nice.

Argyris, C. (1982). *Reasoning, Learning and Action*, San Francisco: Josey-Bass.

Argyris, C. and Schon, D. (1978). *Organizational Learning: A Theory of Action Perspective*, Reading, MA.: Addison-Wesley.

Arora, A., Fosfuri, A. and Gambardella, A. (2000). "Markets for Technology", Conference in Honour of Paul David, Turin, May 20-21.

Arrow, K.J. (1962). "Economic Welfare and the Allocation of Resources for Invention," in Nelson, R.R. (ed) *The Rate and Direction of Inventive Activity*, New York: Wiley.

Arrow, K.J. (1994). "Methodological Individualism and Social Know-ledge", *AEA Paper and Proceedings*, Vol. 84, No. 2, pp. 1-9.

Axelrod, R. (1974). *The Evolution of Cooperation*, New York: Basic Books.

Bacon, F. ([1620] 1960). *Novum Organum*, New York: Liberal Arts Press.

Baldwin, C. and Clark, K. (1997). "Managing in an Age of Modularity", *Harvard Business Review*, July-August.

Barnes, B. (ed) (1972). *Sociology of Science*, New York: Harmondsworth, Penguin.

Beck, U. (1991). *Politik in der Risikogesellschaft*, Frankfurt am Main: Suhrkamp.

Becker, B.E. and Huselid, M.A. (1998). "Human Resources Strategies, Complementarities, and Firm Performance", Working paper, July.

Beckmann, M.J., Johansson, B., Snickers, F. and Thord, R. (eds) (1998). *Knowledge and Networks in a Dynamic Economy*, Berlin: Springer.

Bell, D. (1971). *The Post-Industrial Society*, Boston: Basic Books.

Bennis, W.G. (1971). "Changing Organisations", in Horstein et. al. (eds) *Social Intervention: A Behavioural Science Approach*, New York: Free Press.

Berger, P. and Luckmann, T. (1967). *The Social Construction of Reality*, Garden City, NY.: Doubleday.

Betcherman, G., Leckie, N. and McMullen, K. (2000). "Learning in the Workplace: Training Patterns and Training Activities", in Rubenson, K. and Schuetze, H. G. (eds) *Transition to the Knowledge Society: Policies and Strategies for Individual Participation and Learning*, Vancouver: UBC (Institute for European Studies), pp. 283 - 303.

Betcherman, G., McMullen, K. and Davidman, K. (1998). "Training for the New Economy: A Synthesis Report", Ottawa: Canadian Policy Research Network.

Böhme, G., van den Daele, W. and Hohlfeld, R. (1978). "Finalisierung revisited", in Böhme, G. et al. (eds) *Die gesellschaftliche Orientierung des wissenschaftlichen Fortschritts*, Starn-berger Studien 1. Frankfurt am Main: Suhrkamp, pp. 198-250.

Böhme, G., van den Daele, W. and Krohn, W. (1973). "Finalisierung der Wissenschaft", *Zeitschrift für Soziologie*, Vol. 2, pp. 128-144.

Boisot, M.H.J. (1998). *Knowledge Assets*, Oxford: Oxford University Press.

Boltanski, L. and Chiapello, E. (1999). *Le nouvel âge du capitalisme*, nrf, essais: Gallimard.

Bourdieu, P. (1980). "Le capital social: notes provisoires", *Actes de recherches de sciences sociales*, Vol. 31, No. 2-3.

Braczyk, H.J. et al. (eds) (1998). *Regional Innovation Systems: The Role of Governance in a Globalized World*, London: UCL Press.

Branscomb, L.M., Kodama, F. and Florida, R. (eds) (1999). *Industrializing Knowledge: University-Industry Linkages in Japan and the United States*, Cambridge, MA.: The MIT Press.

Brown, S. and Einsenhardt, K. (1995). "Product Development: Past Research, Present Findings, and Future Directions", *Academy of Management Review*, Vol. 20, No. 2, pp. 343-378.

Brown, J.S. and Duguid, P. (1991). "Organizational Learning and Communities-Of-Practice: Toward a Unified View of Working, Learning and Innovation", *Organization Science*, Vol. 2, No.1, pp. 40–57.

Burns, T. and Stalker, G.M. (1961). *The Management of Innovation*, London: Tavistock.

Bushby, H., Williams, G. and Rogers, A. (1997). "Bodies of Knowledge: Lay and Biomedical Understandings of Musculoskeletal Disorders", in Elston, M.-A. (ed) *The Sociology of Medical Science and Technology*, Oxford: Blackwell, pp. 79-99.

Callon, M. (2001). "The Increasing Involvement of Concerned Groups in R&D Policies: What Lessons for Public Powers?", Ecole des Mines de Paris, Centre de Sociologie, Working Paper.

Carnoy, M., Castells, M., Cohen, S.S. and Cardoso, F.H. (1993). *The Global Economy in the Information Age*, University Park, PA.: Penn State University Press.

Caron, F. (1997). *Les deux révolutions industrielles du XX° siècle*, Albin Michel.

Carter, C. and Williams, B. (1957). *Industry and Technical Progress*, Oxford: Oxford University Press.

Carty, A.J. (2000). In the advertising supplement, *Macleans*, Toronto magazine.

Cavilli-Sforza, L.L. and Feldman, M.W. (1981). *Cultural Transmission and Evolution: A Quantitative Approach*, Princeton, N.J.: Princeton University Press.

Cockburn, I. and Henderson, R. (1998). "Absorptive Capacity, Co-Authoring Behaviour, and the Organization of Research in Drug Discovery", *The Journal of Industrial Economics*, Vol. 46, No. 2, June, pp. 157-182.

Cockburn, I., Henderson, R. and Stern, S. (1999). "Balancing Incentives: The Tension Between Basic and Applied Research", NBER Working Paper, No. W6882, January.

Cohen, D. and Prusak, L. (2001). *In Good Company: How Social Capital Makes Organizations Work*, Cambridge, MA.: Harvard Business School Press.

Cohen, W.M. and Levinthal, D.A. (1989). "Innovation and Learning: The Two Faces of R&D", *The Economic Journal*, Vol. 99, No. 3, September, pp. 569-596.

Cohen, W.M. and Levinthal, D.A. (1990). "Absorptive Capacity: A New Perspective on Learning and Innovation", *Administrative Science Quarterly*, Vol. 35, pp. 128–152.

Coleman, J.S. (2000). "Social Capital in the Creation of Human Capital", in Dasgupta, P. and Serageldin, I. (eds) *Social Capital: A Multifaceted Perspective*, Washington, D.C.: The World Bank.

Collins, H.M. ([1985] 1992). *Changing Order. Replication and Induction in Scientific Practice*, Chicago: University of Chicago Press.

Colm, G. (1934). "Productivity - Statistics", in Seligman, E.R.A. (ed) *Encyclopedia of the Social Sciences*: *Volume Twelve*, New York: Macmillan, pp. 467-472.

Conference Board of Canada (2001). "Collaborating for Innovation: 2nd Annual Innovation Report, based on 1999 Survey of Innovation", Ottawa: Conference Board of Canada.

Cooke, P. et. al. (eds) (2000). *The Governance of Innovation in Europe: Regional Perspectives on Global Competitiveness*, New York: Pinter.

Cortada, J.W. (1998). "Where Did Knowledge Workers Come From?", in Cortada, J.W. (ed) *The Rise of the Knowledge Worker*, Boston: Butterworth-Heinemann.

Coutrot, B. and Paraire, J.L. (1996). "Le développement récent des politiques de motivation des salariés", in D.A.R.E.S. (ed) *Relations professionnelles et négociations d'entreprises en 1992*, Les dossiers de la D.A.R.E.S., No. 3-4, May, pp. 173-178.

Cowan, R., David, P.A. and Foray, D. (2000). "The Explicit Economics of Knowledge Codification and Tacitness", *Industrial and Corporate Change*, Vol. 9., No. 2.

Cowan, R., Fauchart, E., Foray, D. and Gunby, P. (2000). "Technological accidents: learning from disaster", Report for le Programme Crise du CNRS, IMRI, Universite de Paris Dauphine.

DARES, (2000). "La structure des Emplois en 1997", Les dossiers de la DARES, N°3, La Documentation Française.

Dasgupta, P. and Serageldin, I. (eds) (2000). "Social Capital: A Multifaceted Perspective", Washington, D.C.: The World Bank.

Davenport, T.H. and Prusak, L. (1998). *Working Knowledge, How Organizations Manage What They Know*, Cambridge, MA.: Harvard Business School Press.

David, P.A. (1997). "The Knowledge Factory: A Survey of Universities", *The Economist*, October 4, 1997.

David, P.A. (2001). "Digital Digital technologies, research collaborations and the extension of protection for intellectual property in science: will building 'good fences' really make 'good neighbors'"?, STRATA-ETAN Workshop, Brussels, January, 2001.

David, P.A. and Foray, D. (1995). "Accessing and Expanding the Science and Technology Knowledge", STI Review, No. 16.

David, P.A. and Wright, G. (1991). "Resource abundance and American economic leadership", Stanford University, Discussion Paper, No. 267, CEPR.

de la Mothe, J. (1992a). "Competing Through Innovation", in Science Council of Canada, *Reaching for Tomorrow: Science and Technology Policy 1991*, Ottawa: Supply and Services Canada.

de la Mothe, J. (1992b). "A Dollar Short and a Day Late: A Note on the Demise of the Science Council of Canada", *Queen's Quarterly*, December.

de la Mothe, J. (1994). "Canada and the National System of Innovation", in Industry Canada, *Science and Technology Strategy Resource Book, Volume II*, Ottawa: Industry Canada.

de la Mothe, J. (1999). "Empowering Information and Networks Through Adaptive Public Policies", in de la Mothe, J. and Paquet, G. (eds) *Information, Innovation and Impacts*, Boston: Kluwer Academic Publishers.

de la Mothe, J. and Paquet, G. (1994). "Circumstantial Evidence: A Note on Science Policy in Canada", *Science and Public Policy*, December.

de la Mothe, J. and Paquet, G. (1996a). "The Technology-Trade Nexus", *Technology in Society*.

de la Mothe, J. and Paquet, G. (eds.) (1996b). *Evolutionary Economics and the New International Political Economy*, London: Pinter.

de la Mothe, J. and Paquet, G. (eds) (1998). *Local and Regional Systems of Innovation*, Boston: Kluwer Academic Publishers.

de la Mothe, J. and Paquet, G. (eds) (2000). *Information, Innovation and Impacts*, Boston: Kluwer Academic Publishers.

De Meyer A., Dubuisson S. and Le Bas C., (1999). (Title??) in Foray, D. and Mairesse, J. (eds) *Innovations et performances, approaches interdisciplinaires*, Paris: Editions de l'École des Hautes Études en Sciences Sociales.

Denning, S. (2000). The Springboard: How Storytelling Ignites Action in Knowledge-Era Organizations, Woburn, MA.: Butterworth-Heinemann.

Dery, D. (1986). "Knowledge and Organisations", *Policy Studies Review*, Vol. 6, No. 1.

Dewey, J. (1929). *The Quest for Certainty: A Study of the Relations of Knowledge to Action*, New York: Perigee.

Dogson, M. (1991). *The Management of Technological Learning: Lessons from a Biotechnology Company*, Berlin: Walter & Gruyter.

Dosi, G. (1982). "Technological Paradigms and Technological Trajectories", *Research Policy*, Vol. 11, No. 3.

Dosi, G. (1996). "The Contribution of Economic Theory to the Understanding of a Knowledge-Based Economy" in OECD, *Employment and Growth in the Knowledge-Based Economy*, Paris: OECD.

Dosi, G., Teece, D.J. and Winter, S. (1992). "Toward a Theory of Corporate Coherence: Preliminary Remarks", in Dosi, G. and Toninelli, P. (eds) *Technology and Enterprise in a Historical Perspective*, Oxford: Clarendon Press.

Doutriaux, J. and Barker, M. (1995). "The University–Industry Relationship in Science and Technology", Occasional Paper No. 11, Ottawa: Industry Canada.

Drucker, P. (1993). *The Post-Capitalist Society*, Los Angeles: Beacon.

Duguet, E. and Greenan, N. (1997). "Le biais technologique: une analyse économétrique sur les données individuelles", *Revue Economique*, Vol. 48, No. 5, September, pp. 1061-1089.

Edquist, C. (1997). "Systems of Innovation Approaches: Their Emergence and Characteristics", in Edquist, C. (ed) *Systems of Innovation: Technologies, Institutions and Organizations*, London: Pinter.

Edquist, C. (ed) (1997). *Systems of Innovation: Technologies, Institutions and Organizations*, London: Pinter.

Edquist, C. and Hommen, I. (1999). "Systems of Innovation: Theory and Policy for the Demand Side", *Technology in Society*, Vol. 21, pp. 63-79.

Edquist, C. and Riddell, W.C. (2000). "The Role of Knowledge and Innovation for Economic Growth and Employment in the Information and Communication Technology Area", in Rubenson, K. and Schuetze, H.G. (eds), *Transition to the Knowledge Society: Policies and Strategies for Individual Participation and Learning*, Vancouver: UBC (Institute for European Studies), pp. 3-32.

Elias, N. (1991). *The Symbol Theory*, London: Sage.

Engels, F. (1956). "Umrisse zu einer Kritik der Nationalökonomie" in Marx and Engels, *Werke, Volume 1*, Berlin: Dietz.

Epstein, S.R. (1998). "Craft Guilds, Apprenticeship, and Technological Change in Preindustrial Europe", *The Journal of Economic History*, Vol. 58, No. 3.

Etheridge, L. (1985). "Government Learning: An Overview", in Long, S. (ed) *The Handbook of Political Behaviour, Volume 2*, New York: Plenum.

Etzkowitz, H. (1999). "Bridging the Gap: The Evolution of University-Industry Links in the United States," Branscomb, L.M., Kodama, F. and Florida, R. (eds) *Industrializing Knowledge: University-Industry Linkages in Japan and the United States*, Cambridge, MA.: The MIT Press.

Eurostat, (1997). *The First Community Innovation Survey*, Brussels: European Commission.

Eurostat, (1999). "Community Innovation Survey 1997/1998", *Statistics in Focus, Research and Development*, No. 2, 99.

Evangelista, R., Sandven, T., Sirilli, G. and Smith, K. (1997). "Innovation Expenditures in European Industry", Report to the European Commission, DG-XIII-C, European Innovation Monitoring Initiative. p. 46

Faulkner, W. and Senker, J. (1995). *Knowledge Frontiers: Public Sector Research and Industrial Innovation in Biotechnology, Engineering Ceramics, and Parallel Computing*, Oxford: Clarendon Press.

Fleck, L. ([1935] 1975). *The Genesis and Development of a Scientific Fact*, Chicago: University of Chicago Press.

Florida, R. (1998). "Calibrating the Learning Region", in de la Mothe, J. and Paquet, G. (eds), *Local and Regional Systems of Innovation*, Boston: Kluwer Academic Publishers.

Foray, D. (1998). "Comment mesurer l'économie de la connaissance? Un cadre analytique", Paris: Université Paris Dauphine, Working Paper IMRI – WP 98/10.

Foray, D. (2000). "Characterising the Knowledge Base: Available and Missing Indicators", in OECD, (2000) Knowledge *Management in the Learning Society*, Paris: OECD.

Foray, D. and Lundvall, B.-A. (1996). "The Knowledge-Based Economy: from the Economics of Knowledge to the Learning Economy", in OECD, *Employment and Growth in the Knowledge-Based Economy*, OECD Documents, Paris: OECD.

Foray, D. and Mairesse, J. (eds) (1999). "Innovations et performances: approches interdisciplinaires", Paris: Éditions de l'école des hautes études en sciences sociales.

Foray, D. and Steinmueller, W.E. (2001). "Codification and the Domestication of Tacit Knowledge", Paper presented at the Conference in honour of R.Nelson and S.Winter, Aalborg.

Foss, N.J. (1997). *Resources, Firms, and Strategies: A Reader in the Resource-Based Perspective*, Oxford: Oxford University Press.

Fountain, J.E. (1998). "Social Capital: Its Relationship to Innovation in Science and Technology", *Science and Public Policy*, Vol. 25, No. 3, pp. 103-115.

Fourastié, J. ([1951] 1954). *The Causes of Wealth.*, Glencoe, IL: Free Press.

François, J.P. (1997). "Les compétences pour innover dans l'industrie française : premiers résultats", colloquim on: The développement of practical tools for improving the innovation performance of firms, OECD, June, 30, - July 1, Paris.

François, J.P., Goux, D., Guellec, D., Kabla, I., and Templé, P. (1999). Le développement d'un outil pour mesurer les compétences : l'enquête "Compétences pour Innover", in OECD work on measuring intangible investment, OCDE, Paris. In Mairesse and Foray ed., 1999.

Francoz, D. (2000). "Measuring R&D in R&D an Innovation Surveys: Analysis of Causes of Divergence in Nine OECD Countries", Working Party of National Experts on Science and Technology Indicators, OECD, Paris, 26, May.

Freeman, C. and Soete, L. (1997). *The Economics of Industrial Innovation*, London: Pinter.

Fukuyama, F. (1995). *Trust*, London: Hamish Hamilton.

Galbraith, J.K. (1967). *The New Industrial State*, Boston: Houghton Mifflin.

Gershuny, J. and Miles, I. (1982). *The New Service Economy*, London: Blackwell.

Geuss, R. (1981). *The Idea of Critical Theory*, Cambridge: Cambridge University Press.

Gibbons, M. (1995). "Technology and the Economy: A Review of Some Critical Relationships", Industry Canada Occasional Paper, Number 12, Ottawa: Industry Canada, December.

Gibbons, M., Limoges, C., Nowatny, H., Schwartzman, S., Scott, P. and Trow, M. (1994). *The New Production of Knowledge*, London: Sage.

Gimpel, J. (1975). *La révolution industrielle du Moyen Âge*, Seuil, Point.

Gingras, Y., Masse, P. and Roy, R. (2000). "The Changing Skill Structure of Employment in Canada", in Rubenson, K. and Schuetze, H. G. (eds) *Transition to the knowledge society: Policies and strategies for individual participation and learning*, Vancouver: UBC (Institute for European Studies)

Gjerding, A.N. (1996). "Organizational Innovation in the Danish Private Business Sector", DRUID working paper, No. 96-16.

Goody, J. (1977). *The Domestication of the Savage Mind*, Cambridge: Cambridge University Press.

Greenan, N. (1995). "L'organisation de la production dans l'industrie", Le 4 pages, SESSI éditeur, No. 43, Janvier.

Greenan, N. (1996). "Innovation technologique, changements organisationnels et évolution des compétences", *Economie et Statistiques*, No. 298, pp. 15-33.

Greenan, N. and Guellec, D. (1998). "Firm Organization, Technology and Performance: An Empirical Study", *Economics of Information and New Technologies*, Vol. 6, pp. 313-347.

Greenan, N., Mairesse J. and Topiol-Bensaid A. (2001). "Information Technology and Research and Development Impacts on Productivity and Skills: Looking for Correlations on French Firm Level Data", NBER Working Paper, No. W8075, January.

Greenan, N.and Mairesse, J. (1999). "Organizational Change in French Manufacturing: What Do We Learn From Firm Representatives and From Their Employees?", NBER Working Paper, No. W7285, August.

Grindley, P. and Teece, D. (1997). "Managing Intellectual Capital: Licensing and Cross-Licensing in Semiconductors and Electronics", *California Management Review*, Vol. 39, No. 2.

Haas, E. (1990). *When Knowledge is Power*, Berkeley: University of California Press.

Habermas, J. (1981). *Theorie des kommunikativen Handelns: Volume Two.* Frankfurt am Main: Suhrkamp.

Hall, P.A. (ed) (1989). *The Political Power of Economic Ideas*, Princeton: Princeton University Press.

Hansen, M. (1999). "The Search-Transfer Problem: The Role of Weak Ties in Sharing Knowledge Across Organisation Subunits", *Administrative Science Quarterly*, Vol. 44.

Hansen, W.K. (1999). "Analyse du redéploiement des travailleurs du domaine de la science et de la technologie dans l'économie canadienne", Document de travail, Statistics Canada, N°88-006XIB-99003.

Hargreaves, D. (2000). "The Production, Mediation and Use of Professional Knowledge Among Teachers and Doctors: A Comparative Analysis", in OECD, *Knowledge Management in Learning Society*, Paris: OECD.

Hayek, F.A. ([1945] 1948). "The Use of Knowledge in Society", in Hayek, F.A., *Individualism and Economic Order*, Chicago: University of Chicago Press.

Hayek, F.A. (1937). "Economics and Knowledge", *Economica*, Vol. 33.

Helmstädter, E. (2000). "Wissensteilung", Graue Reihe Nr. 12 des Wissenschaftszentrums Nordrhein-Westfalen, Institut Arbeit und Technik, Gelsenkirchen.

Hicks, D. (1995). "Published Papers, Tacit Competences and Corporate Management of the Public/Private Character of Knowledge", *Industrial and Corporate Change*, Vol. 4, No. 2.

Hicks, D., Breitzman T., Olivastro D., and Hamilton K. (2001). "The Changing Composition of Innovative Activity in the U.S.: A Portrait Based on Patent Analysis", *Research Policy*, (forthcoming).

Hippel, E. von (1987). "Cooperation Between Rivals: Informal Know-How Trading", *Research Policy*, Vol. 16, No. 6, December, pp 291-302.

Hippel, E. von (1988). *The Sources of Innovation*, Oxford: Oxford University Press.

Hippel, E. von (1991). "The Impact of 'Sticky Information' on Innovation and Problem-Solving", Sloan School of Management, MIT, Working Papers BPS 33147 (revised).

Hippel, E. von (1994). "'Sticky Information' and the Locus of Problem Solving: Implications for Innovation", *Management Science*, Vol. 40, pp. 429-439.

Hippel, E. von (1994). "Sticky Information and the Locus of Problem Solving: Implications for Innovation", *Management Science*, Vol. 40, No. 4.

Hippel, E. von and Tyre, M. (1995). "How Learning By Doing is Done: Problem Identification in Novel Process Equipment", *Research Policy*, Vol. 24.

Holbrook, J.A. and Wolfe, D.A. (eds) (2000). *Innovation, Institution and Territory. Regional Innovation Systems in Canada*, Kingston: McGill-Queen's University Press.

Hölmstrom, B. (1989). "Agency Costs and Innovation", *Journal of Economic Behaviour and Organization*, Vol. 12, No. 3, pp. 305-327.

Honig-Haftel, S. and Martin, L. (1993). "The Effectiveness of Reward Systems on Innovative Output: An Empirical Analysis", *Small Business Economics*, Vol. 5, No. 4, December, pp. 261-69.

Huselid, M.A. (1995). "The Impact of Human Resource Management Practices on Turnover, Productivity, and Corporate Financial Performance", *Academy of Management Journal*, Vol. 38, pp. 635-672.

Iansiti, M. (1995). "From Technological Potential to Product Performance: an Empirical Analysis", *Research Policy*, Vol. 26, No. 3, pp. 345-367.

Ichniowski, C., Shaw, K. and Prennushi, G. (1997). "The Effects of Human Resource Management Practices on Productivity: A Study of Steel Finishing Lines", *American Economic Review*, Vol. 87, No. 3, June, pp. 291-313.

Irwin, A. and Wynne, B. (1996). "Conclusions", Irwin, A. and Wynne, B. (eds) *Misunderstanding Science? The Public Reconstruction of Science and Technology*, Cambridge: Cambridge University Press.

Jacobsson, S. and Oskarsson, C. (1995). "Educational Statistics as an Indicator of Technological Activity", *Research Policy*, Vol. 24, No. 1, pp. 127-136.

Jaffe, A.B. (1986). "Technological Opportunity and Spillovers of R&D: Evidence from Firms' Patents, Profits and Market Value", *American Economic Review*, Vol. 76, No. 5, pp.984-1001.

Jervis, R. (1976). *Perception and Misperception in International Politics*, Princeton: Princeton University Press.

Jervis, R. (1997). *System Effects: Complexity in Political and Social Life*, Princeton: Princeton University Press.

Jewkes, J., Sawers, D. and Stillerman, R. (1962). *The Sources of Invention*, London: Macmillian.

Johnson, B. (1995). "Institutional Learning", in Lundvall, B.-A. (ed) *National Systems of Innovation*, London: Pinter.

Johnson, B. and Lundvall, B.-A. (2001). "Why All This Fuss About Codified and Tacit Knowledge?", Paper presented to the DRUID Winter Conference, Korsor, Denmark, January 18-20.

Kanter, R.M. (1999). "Change Is Everyone's Job: Managing the Extended Enterprise in a Globally Connected World", *Organizational Dynamics*, Vol. 28, No. 1, pp. 7-23, Summer.

Keating, D. and Hertzman, C. (eds) (1999). *Developmental Health and the Wealth of Nations: Social, Biological and Educational Dynamics*, New York: Guildford.

Keohane, R. and Nye, J. (1987). "Power and Interdependence Re-Visited", *International Organisation*, Vol. 4, No. 4.

Kinkel, S. and Wengel, J. (1997). "New Production Concepts: One Discussion Does Not Make a Summer", *News from the Manufacturing Innovation Survey*, No. 4, I.S.I., Karlsruhe.

Kleinknecht, A. (2001). "How Misleading are Innovation Indicators, Workshop on Innovation, Technological Change and Growth in Knowledge-Based and Service Intense Economies", The Royal Institute of Technology, Stockholm, February 1-2.

Kleinknecht, A. and Reijnen, O.N. (1991). "More Evidence on the Undercounting of Small Firm R&D", *Research Policy*, Vol. 20, pp.579-587.

Kleinknecht, A. and Wunderink, S. (1999). "Sales of Innovative Products: a Problematic Innovation Indicator?", RIBES Discussion Paper, No. 9962.

Kleinknecht, A., Poot, T.P. and Reijnen J.O.N. (1991). "Formal and Informal R&D and Firm Size: Survey Results from the Netherlands", in Acs, Z.J. and Audretsch, D.B. (eds) *Innovation and Technical Change, an International Comparison*, Harvester Wheatsheaf, pp.84-108.

Klevorick, A.K., Levin, R., Nelson, R.R. and Winter, S. (1995). "On the Sources and Significance of Inter-Industry Differences in Technological Opportunities", *Research Policy*, Vol. 24, pp. 185-205.

Kline, S.J. and Rosenberg, N. (1986). "An Overview of Innovation," in Landau, R. and Rosenberg, N. (eds) *The Positive Sum Strategy: Harnessing Technology for Economic Growth*, Washington, DC: National Academy Press.

Knack, S. and Keefer, P. (1997). "Does Social Capital Have Economic Payoff? A Cross-Country Investigation", *The Quarterly Journal of Economics*, pp. 1251-1288.

Kodama, F. (1991). *New Patterns of Innovation*, Cambridge MA.: Harvard Business School Press.

Krogh, G. von, Ichijo, K. and Nonaka, I. (2000). *Enabling Knowledge Creation*, Oxford: Oxford University Press.

Kuhn, T. (1970). *The Structure of Scientific Revolution*, Chicago: University of Chicago Press.

Kusunoki, K., Nonaka I. and Nagata, A. (1998). "Organizational Capabilities in Product Development of Japanese Firms: A Conceptual Framework and Empirical Findings", *Organization Science*, Vol. 9., No. 6, November-December, pp. 699-718.

Kwašnicki, W. (1996). *Knowledge, Innovation and Economy*, Cheltenham: Edward Elgar.

Lacourrège, A. and Jacquin, Y. (2000). "Les chercheurs en entreprise : évolution sur longue périodeet situation en 1997", Note d'information, 00.10, mars, (http://www.education.gouv.fr/dpd/ni.htm).

Lakatos, I. and Musgrave, A. (eds) (1970). *Criticism and the Growth of Knowledge*, Cambridge: Cambridge University Press.

Lakhani, K. and Hippel, E. von (2000). "How Open Source Software Works: 'Free' User-to-User Assistance", MIT Sloan School of Management Working Paper, 4117.

Lam, A. (1998). "Tacit Knowledge, Organisational Learning and Societal Institutions", Paper presented to the XIV. World Congress of Sociology, Montreal, Canada, July 26-August 1.

Landry, R. (1996). "Enquête sur les capacités technologiques et les besoins des entreprises innovantes de la région Chaudière-Appalaches", Rapport préparé pour l'Antenne technologique Chaudière-Appalaches.

Landry, R. (1998). "L'innovation de produits et de procédés de fabrication dans les entreprises manufacturières de l'Estrie: état de la situation et pistes d'action", Rapport préparé pour le Groupe d'action technologique de l'Estrie [GATE].

Landry, R. and Amara, N. (1998). "The Chaudière-Appalaches System of Industrial Innovation", in de la Mothe, J. and Paquet, G. (eds) *Local and Regional Systems of Innovation*, Boston: Kluwer Academic Publishers.

Landry, R. and Frechette, P. (1998). "L'innovation dans les entreprises manufacturières de la région Québec-Chaudière-Appalaches", Rapport préparé pour le Groupe d'action technologique et industriel de la région de Québec-Chaudière-Appalaches [GATICA].

Landry, R., Amara, N. and Lamari, M. (2000). "Innovation et technologies avancées dans les entreprises manufacturières de la Montérégie: bilan et pistes d'action", Étude réalisée pour Valo Tech Inc.

Landry, R., Lamari, M. and Amara, N. (2000). "Bilan technologique de la Mauricie: analyse des résultats de l'enquête quantitative auprès des entreprises manufacturières", Étude réalisée pour la Technopole de la Vallée du St-Maurice.

Landry, R., Lamari, M. and Amara, N. (2001). "Does Social Capital Determine Innovation? To What Extent ?", *Technological Forecasting and Social Change*, (forthcoming).

Larson, D. (1985). *Origins of Containment: A Psychological Explanation*, Princeton: Princeton University Press.

Laursen, K. and Foss, N.J. (2000). "New HRM Practices, Complementarities, and the Impact on Innovation Performance", Working Paper, Department of Industrial Economics and Strategy, Copenhagen Business School, 3rd draft.

Lay, G. (1997). "Process Innovations as the Key to Innovative Products, News from the Manufacturing Innovation Survey", No. 7, I.S.I., Karlsruhe.

Le Bas, C., Picard, F. and Suchecki, B. (1998). "Innovation technologique, comportement de réseaux et performances: une analyse sur données individuelles", *Revue d'économie politique*, Vol. 108, No. 5, pp. 625-644.

Le Masson, P. (2000). "La gestion de la connaissance pour l'innovation, revue de la littérature", Rapport pour l'ANRT, Centre de Gestion Scientifique, École des Mines de Paris

Leeuw, F.L., Rist, R.C. and Sonnichesen, R.C. (eds) (2000). *Can Governments Learn?: Comparative Perspectives on Evaluation and Organizational Learning*, New York: Transaction.

Lengrand, L. and Chatrie, I. (1999). "Business Networks and the Knowledge-Driven Economy", Brussels: European Commission.

Leonard, D. (1995). *Wellsprings of Knowledge*, Cambridge, MA.: Harvard Business School Press.

Lesser, E.L. (2000). *Knowledge and Social Capital: Foundations and applications*, Boston: Butterworth Heinemann.

Levitt, B. and March, J. (1996). "Organizational Learning, in Organizational Learning", in Cohen, M.D. and Sproull, L.S. (eds) *Organization Science*, Sage Publications.

Lhuillery, S. (2000). "Managing Innovation Through Complementarities and Specificities of Organizational Practices: An Empirical Investigation", Working Paper, Paris: University Paris Nord.

Lhuillery, S. and Templé, P. (1994). "L'organisation de la R&D dans les PMI-PME", *Économie et Statistique*, No. 271-272, 1994, 1/2, INSEE, Paris.

Licht, G., Schnell, W. and Stahl, H. (1995). "Results of the German Innovation Survey", Manheim: ZEW.

Lipsey, R. (2000). "New Growth Theories and Economic Policy for the Knowledge Economy", in Rubenson, K. and Schuetze, H.G. (eds) *Transition to the Knowledge Society: Policies and Strategies for Individual Participation and Learning*, Vancouver: UBC (Institute for European Studies).

Loasby, B.J. (1999). *Knowledge, Institutions and Evolution in Economics*, London: Routledge.

Loasby, B.J. (2000). "Organisation, Competition and the Growth of Knowledge", mimeo.

Lopez, K. (2001). "How to Measure the Value of Knowledge Management", *Knowledge Management Review*, Vol. 2, No. 1., pp. 20-23.

Luhmann, N. ([1992] 1998). *Observations on Modernity*, Stanford, CA: Stanford University Press.

Lundvall, B.-A. (1988). "Innovation as an Interactive Process-from User-Producer Interaction to the National System of Innovation" in Dosi, G. et al. (eds) *Technical Change and Economic Theory,* London: Pinter.

Lundvall, B.-A. (1995). "Introduction", in Lundvall, B.A. (eds), *National Systems of Innovation: Towards a Theory of Innovation and Interactive Learning*, London: Pinter.

Lundvall, B.-A. (ed) (1995). *National Systems of Innovation*, London: Printer.

Lundvall, B-A. (1992). "Explaining Interfirm Cooperation- the Limits of Transaction Cost Approach", in Grabher, G. (ed) *The Embedded Firm: On the Socioeconomics of Industrial Networks*, London: Routledge.

Machlup, F. (1962). "The Production and Distribution of Knowledge in the United States", Princeton: Princeton University Press.

Maillat, D. (1995). "Systèmes territoriaux de production, milieux innovateurs et politiques régionales", Université de Neuchâtel, Suisse, version provisoire non publiée.

Mairesse, J. and Greenan, N. (1999). "Using Employee Level Data in a Firm Level Econometric Study", NBER Working Paper, No. W7028, March.

Mairesse, J. and Mohnen, P. (2001). "To Be or Not To Be Innovate: An Exercise in Measurement", Workshop on Innovation, technological change and growth in knowledge-based and service intense economies, The Royal Institute of Technology, Stockholm, February 1-2, 2001

Malerba, F. (1992). "Learning by Firms and Incremental Technical Change", *The Economic Journal*, Vol. 102, No. 413, pp. 8445-859.

Malerba, F. and Montobbio, F. (2000). "Knowledge Flows, Structure of Innovative Activity and International Specialization", CESPRI Working Paper, No. 119, November 2000.

Marshall, A. ([1890] 1964). *Principles of Economics*, London: Macmillan.

Martin, B. and Salter, A. with Hicks, D., Pavitt, K., Senker, J. Sharp, M. and von Tunzelmann, N. (1996). "The Relationship with Between Publicly Funded Basic Research and Economic Performance: A SPRU Review", Report Prepared for HM Treasury, Brighton: University of Sussex.

Marx, K. (1966). "Theorien über den Mehrwert. Erster Teil", in Marx, and Engels, *Werke*. Volume 26. Berlin: Dietz.

Maskell, P. (1999). "Social Capital, Innovation and Competitiveness", in Baron, S., Field, J. and Schuller, T. (eds) *Social Capital*, Oxford: Oxford University Press.

Mendelson, R. and Pillai, H. (1999). "Information-Age Organizations, Dynamics and Performance", *Journal of Economic Behavior and Organization*, Vol. 38, pp. 253-281.

Miles, I. (2000). "Interactivity and Intangibility: Another Pair of 'I's'", in de la Mothe, J. and Paquet, G. (eds) *Information, Innovation and Impacts*, Boston: Kluwer Academic Publishers.

Miller, R. (1996). *Human Capital Formation*, Paris: OECD.

Mowery, D.C. and Rosenberg, N. (1998). *Paths of Innovation*, Cambridge: Cambridge University Press.

Mowery, D.C. and Rosenberg, N. (1978), "The Influence of Market Demand upon Innovation: A Critical Review of Some Recent Empirical Studies", *Research Policy*, 8, April.

Mulkay, M. (1979). *Science and the Sociology of Knowledge*, London: Allen and Unwin.

Munier, F. (1999). “Les compétences pour innover dans l'industrie, Avantages comparés des tailles d'entreprises: une vérification empirique”, Le 4 pages du SESSI, N°120.

Myers, P. (1996). “Knowledge Management Tools”, Butterworth-Heinemann.

Myers, S. and Marquis, D.G. (1969). “Successful Industrial Innovation”, Washington, DC: National Science Foundation.

N.S.F., (2000). *Annual Report, Science and Engineering Indicators*, Washington: NSF.

Neef, D.G. (1998). *The Knowledge Economy*, Boston: Butterworth-Heinneman.

Neef, D.G., Siesfeld, A. and Cefola, J. (1998). *The Economic Impact of Knowledge*, Boston: Butterworth-Heinneman.

Nelson, R.R. (2000). “Selection Criteria and Selection Processes in Cultural Evolution”, in Ziman, J. (ed) *Technological Innovation as an Evolutionary Process*, Cambridge: Cambridge University Press.

Nelson, R.R, and Winter, S.G. (1982). “An Evolutionary Theory of Economic Change”, Cambridge: Harvard University Press.

Nelson, R.R. (1959). “The Simple Economics of Basic Scientific Research,” *Journal of Political Economy*, Vol. 67, pp. 297-306.

Nelson, R.R. (1996). “Capitalism as an Engine of Progress,” in Nelson, R.R., *The Sources of Economic Growth*, Cambridge, MA.: Harvard University Press.

Nelson, R.R. (ed) (1993). *National Innovation Systems: A Comparative Analysis*, Oxford: Oxford University Press.

Niosi, J. (1993). “National Systems of Innovation in Search of a Workable Concept”, *Technology in Society*, Vol. 15, pp. 207-227.

Nonaka, I. and Takeuchi, H. (1995). *The Knowledge-Creating Company*, New York: Oxford University Press.

Nonaka, I., Toyama, R. and Nagata, A. (2000). “A Firm as a Knowledge-creating Entity: A New Perspective on the Theory of the Firm”, *Industrial and Corporate Change*, Vol. 9, No. 1, pp. 1-20.

Nowotny, H., Scott, P. and Gibbons, M. (2001). *Rethinking Science: Knowledge and Public in an Age of Uncertainty*, Cambridge: Polity (Blackwell).

NUTEK, (1996). *Towards Flexible Organisations*, 225 pages.

OECD, (1993a). *Proposed Standard Practice for Surveys of Research and Experimental Development, The Frascati Manual*, Paris: OECD.

OECD, (1993b). *Small and Medium-Sized Enterprises: Technology and Competitiveness*. Paris: OECD.

OECD, (1995a). "The Measurement of Scientific and Technical Activities", in OECD, *Manual on the Measurement of Human Resources Devoted to S&T, Canberra Manual*, Paris: OECD, OECD/GD(95)77.

OECD, (1995b). *Boosting Business Advisory Services*. Paris: OECD.

OECD, (1995c). *Literacy, Economy and Society*, Paris: OECD.

OECD, (1996a). "Special Theme: The Knowledge–Based Economy, Science, Technology and Industry Outlook 1996", Paris: OECD.

OECD, (1996b). "OECD Proposed Guidelines for Collecting and Interpreting Technological Innovation Data, 'Oslo Manual'", Paris: OECD.

OECD, (1997). *Literacy Skills for the Knowledge Society*, Paris: OECD.

OECD, (1999a). *Unlocking Knowledge in the Learning Society*, Paris: OECD.

OECD, (1999b). *The Knowledge-Based Economy: A Set of Facts and Figures*, Paris: OECD.

OECD, (2000a). "High-Level Forum on Knowledge Management: The Challenge for Firms and Organisations", September 21-22, 2000, Ottawa, Canada.

OECD, (2000b). "The Measurement of R&D in Services", OECD, June - DSTI/EAS/STP/NESTI/NESTI(2000) 11 Rev 1

OECD, (2000c). *Knowledge Management in the Learning Society*, Paris: OECD.

OECD, (2000d). *The OECD Small and Medium Enterprise Outlook*. Paris: OECD.

Parson, E.A. and Clark, W.C. (1993). "Sustainable Development as Social Learning: Theoretical Perspectives and Practical Challenges for the Design of a Research Program", *Faculty Research Working Paper Series*, Cambridge MA: John F. Kennedy School of Government, Harvard University, R93-37.

Patel, P. and Pavitt, K. (1994). "National Innovation Systems: Why They Are Important, and How They Might Be Measured and Compared", *Economics of Innovation and New Technology*, 3, pp. 77-95.

Pavitt, K. (1991). "What Makes Basic Research Economically Useful?", *Research Policy*, Vol. 20, pp. 109-119.

Perkins, D. (2000). "The Evolution of Adaptive Form", in Ziman, J. (ed) *Technological Innovation as an Evolutionary Process*, Cambridge: Cambridge University Press.

Polanyi, M. (1958). *Personal Knowledge: Towards a Post-Critical Philosophy*, London: Routledge & Kegan Paul.

Polanyi, M. (1967). *The Tacit Dimension*, New York: Doubleday.

Popper, K.R. (1973). *Objective Knowledge: An Evolutionary Approach*. Oxford: Oxford University Press.

Porta, P.L., Scazzieri, R. and Skinner, A. (eds) (2001). *Knowledge, Social Institutions and the Division of Labour*, London: Edward Elgar.

Porter, M. (1999). "Clusters and the New Economics of Competition", *Harvard Business Review*, Dec, pp. 77-90.

Porter, M. (2000). "Location, Competition and Economic Development: Local Clusters in a Global Economy", *Economic Development Quarterly*, Vol. 14, No. 1, pp. 15-34.

Portes, A. (1998). "Social Capital: Its Origins and Applications in Modern Sociology", *Annual Review of Sociology*, Vol. 24.

Prusak, L. (ed) (1997). *Knowledge in Organizations*, Boston, MA.: Butterworth-Heinemann.

Putnam, R. (1993). *Making Democracy Work*, Princeton, N.J.: Princeton University Press.

Raz, J. ([1975] 1990). *Practical Reasons and Norms*, Princeton, N.J.: Princeton University Press.

Reich, R. (1992). *The Work of Nations*, New York: Knopf.

Rose, R. (1991). "What Is Lesson Drawing?", *Journal of Public Policy*, Vol. 11, No. 1.

Rosenberg, N. (1982). *Inside the Black Box: Technology and Economics*, New York: Cambridge University Press.

Rosenberg, N. (1985). "The Commercial Exploitation of Science by American Industry", in Clark, K.B., Hayes, R.H. and Lorenz, C. (eds), *The Uneasy Alliance: Managing the Productivity-Technology Dilemma*, Boston, MA.: Harvard Business School Press.

Rosenberg, N. (1990). "Why Do Firms Do Basic Research?", *Research Policy*, Vol. 19, pp. 165-174.

Rubenson, K. and Schuetze, H.G. (1996). "Learning At and Through the Workplace - A Review of Participation and Adult Learning Theory", in (author??) *What Makes Workers Learn? – The Role of Incentives in Workplace Education and Training*, Cresskill, N.J.: Hampton Press.

Rubin, M. and Huber, M.T. (1986). *The Knowledge Industry in the United States: 1960-1980*, Princeton: Princeton University Press.

Ruggles, R.L. (ed) (1997). *Knowledge Management Tools*, Boston: Butterworth-Heinemann.

Rugman, D. (1998). *Economics and the Philosophy of Science*, Oxford: Oxford University Press.

Ryan, M.P. (1998). *Knowledge Diplomacy: Global Competition and the Politics of Intellectual Property*, Washington: Brookings.

Rycroft, R. and Kash, D. (1999). *The Complexity Challenge: Technological Innovation for the 21st Century*, London: Pinter.

Saint-Paul, G. (2001). "Information Technology and the Knowledge Elites", Discussion Paper Series Number 2786, Centre for Economic Policy Research, Stanford University.

Santarelli, E. and Sterlacchini, A. (1990). "Innovation, Formal vs. Informal R&D and Firm Size: Some Evidence From Italian Manufacturing Firms", *Small Business Economics*, Vol. 2, No. 3.

Saxenian, A. (1998). Regional Systems of Innovation and the Blurred Firm, in de la Mothe, J. and Paquet, G. (eds) *Local and Regional Systems of Innovation,* Boston: Kluwer Academic Publishers.

Schmookler, J. (1966). "Invention and Economic Growth", Cambridge, MA.: Harvard University Press.

Schreiber, G., Ackkermans, H., Anjewierden, A., de Hoog, R., Shadbolt, N. Van de Velde, W. and Wielinga, B. (2001). *Knowledge Engineering and Management: The Common KADS Methodology*, Cambridge MA.: MIT Press.

Schuetze, H.G. (1998). "How Do Small Firms Innovate in B.C.?", in de la Mothe, J. and Paquet, G. (eds) *Local and Regional Systems of Innovation*, Boston: Kluwer Academic Publishers.

Schuetze, H.G. (2001). "Innovation, Small Firms and Their Environment – A Study of the Secondary Wood Sector in Four Regions in British Columbia", Vancouver: UBC (Centre for Policy Studies in Higher Education and Training).

Schuetze, H.G., Hommen, L., Best, A. and Taylor, R. (1997). "Innovations, Skills and Learning - A Study of Human Resources Management in Small and Medium Enterprises in British Columbia", Vancouver: University of British Columbia (Centre for Policy Studies in Education).

Schumpeter, J. A. ([1911] 1934). *The Theory of Economic Development*, Cambridge, MA.: Harvard University Press.

Senior, N.W. (1850). *Political Economy*, London: J.J. Griffin.

Senker, J. (1995). "Tacit Knowledge and Models of Innovation", *Industrial and Corporate Change*, Vol. 4, No. 2, pp. 425-447.

Shapiro, C. and Varian, H. (1998). *Information Rules*, Cambridge MA: Harvard Business School Press.

Shattuck, R. (1996). *Forbidden Knowledge: From Prometheus to Pornography*. San Diego: Harcourt Brace & Company.

Simmel, G. ([1890] 1989). "Über sociale Differenzierung", in Simmel, G. Aufsätze 1987-1980. Über sociale Differenzierung. Die Probleme der Geschichtsphilosophie (1892). Gesamtausgabe Volume 2. Frankfurt am Main: Suhrkamp.

Simon, H.A. (1982). *Models of Bounded Rationality: Behavioural Economics and Business Organisation, Vol. 2*, Cambridge, MA.: The MIT Press.

Simon, H.A. (1991). "Bounded Rationality and Organizational Learning", *Organization Science*, Vol. 2, No. 1, pp. 125 – 134.

Smith, A. ([1776] 1909). *An Inquiry into the Nature and Causes of the Wealth of Nations*, New York: P.F. Collier and Sons.

Smith, K. (2000). "Innovation Indicators and the Knowledge Economy: Concepts, Results and Policy Challenges", Paper presented to Conference on Innovation and Enterprise Creation: statistics and indicators Sophia Antipolis, France, 23-24 November, 16.

Statistics Canada and Human Resources Development Canada, (1999). "South of the Border: Graduates from the Class of '95 Who Moved to the United States", Ottawa: Statistics Canada, Cat. No. 81-587-XPB.

Statistics Canada, (1998a). "Knowledge Flows In Canada as Measured by Bibliometrics", Ottawa: Statistics Canada, Cat. No. 88F0006XIB98010.

Statistics Canada, (1999a). "An Analysis of Science and Technology Workers: Deployment in the Canadian Economy", Ottawa: Statistics Canada, Cat. No. 88-006XIB-99003.

Statistics Canada, (1999b). "Coverage" (reference products: technical reports: 1996 Census of Population), Ottawa: Statistics Canada, Cat. No. 92-370-XPB.

Statistics Canada, (1999c). "Technology Adoption in Canadian Manufacturing: Survey of Advanced Technology in Canadian Manufacturing", Ottawa: Statistics Canada, Cat. No. 88F0006XIB99005.

Statistics Canada, (2000a). "Survey of Intellectual Property Commercialization in the Higher Education Sector, 1999", Ottawa: Statistics Canada, Cat. No. 88F0006XIB-00001.

Statistics Canada, (2000b). "Workplace and Employee Survey, 1999". *The Daily*, February 19, 2001, Ottawa: Statistics Canada.

Statistics Canada, (2000c). "Mapping Innovation and Connectedness, [in] Innovation Analysis Bulletin", Ottawa: Statistics Canada, Vol. 2, No. 3, Cat. No. 88-003-XIE.

Statistics Canada, (2000d). "Information and Communications Technologies and Electronic Commerce in Canadian Industry", Ottawa: Statistics Canada, Cat. No. 88F0006XIB00004.

Statistics Canada, (2001a). "Innovation Partners of Canadian Manufacturing Firms, [in] Innovation Analysis Bulletin", Ottawa: Statistics Canada, Vol. 3, No. 1, Cat. No. 88-003-XIE.

Statistics Canada, (2001b). "Employee Involvement Practices in Canadian Workplaces", [by] Sacha Senécal, unpublished Data Interpretation Workshop report (January 8 to February 16, 2001), Ottawa, Canada.

Statistics Canada, (2001c). "Estimation of Research and Development Expenditures in the Higher Education Sector, 1998-99", Ottawa: Statistics Canada, Cat. No. 88F0006XIB01002.

Statistics Canada, (2001d). "Innovation, Advanced .Technologies and Practices in the Construction and Related Industries: National Estimates", Ottawa: Statistics Canada, Cat. No. 88F000601004.

Stehr, N. (1992). *Practical Knowledge: Applying the Social Sciences.* London: Sage.

Stehr, N. (2001a). *Knowledge and Economic Conduct: The Social Foundations of the Modern Economy.* Toronto: University of Toronto Press.

Stehr, N. (2001b). "Economy and Ecology in an Era of Knowledge-Based Economies", *Current Sociology*, (in print).

Stehr, N. and Grundmann, R. (2001). "The Authority of Complexity", *British Journal of Sociology*, (in print).

Steinmueller, W.E. (2000a). "Knowledge and Learning in the Information Age", Summer 2000 Conference of DRUID, Aalborg.

Steinmueller, W.E. (2000b). "Learning in the Knowledge-Based Economy: the Future as Viewed from the Past", Conference in Honour of Paul David, Turin, May 20-21.

Stevens, C. (1998). "The Knowledge Driven Economy", in Neef, D. (ed), *The Knowledge Economy*, Boston: Butterworth-Heinnemann.

Stichweh, R. (1999). "Globalisierung von Wirtschaft und Wissenschaft: Produktion und Transfer wissenschaftlichen Wissens in zwei Funktions-sytemen der modernen Gesellschaft", *Soziale Systeme*, Vol. 5, pp. 27-39.

Stigler, G.J. (1961). "The Economics of Information", *The Journal of Political Economy*, Vol. 69, pp. 213-225.

Stokes, D. (1997). *Pasteur's Quadrant: Basic Science and Technological Innovation*, Washington DC.: Brookings Institute.

Storper, M. (1997). *The Regional World: Territorial Development in a Global Economy*, New York: The Guilford Press.

Sveiby, K.E. and Lloyd, T. (1987). *Managing Knowhow*, Bloomsbury.

Sweeney, G. (ed) (1985). *Innovation Policies: An International Perspective*, London: Pinter.

Teece, D.J. (1988). "Research Directions for Knowledge Management", *California Management Review*, Vol. 40, No. 3, Spring, pp. 289-292.

Teece, D.J. (1998). "Capturing Value from Knowledge Assets", *California Management Review*, Vol. 40, No. 3.

Thompson, V.A. (1969). *Bureaucracy and Innovation*, Birmingham: University of Alabama Press.

Toulmin, S. (1961). *Foresight and Understanding*, New York: Harper and Row.

Toulmin, S. (1972). *Human Understanding*, Princeton, N.J.: Princeton University Press.

Tyre, M. and Hippel, E. von (1997). "The Situated Nature of Adaptive Learning in Organizations", *Organization Science*, Vol. 8, No. 1.

Uphoff, N. (2000). "Understanding Social Capital: Learning from the Analysis and Experience of Participation", in Dasgupta, P. and Serageldin, I. (ed) *Social Capital: A Multifaceted Perspective*, Washington, D.C.: The World Bank.

Veblen, T. ([1908] 1919). "On the Nature of Capital", in Veblen, T., *The Place of Science in Modern Civilization and Other Essays*. New York: Viking.

Vickery, G. (1999). "Organizing for Information and Innovation", in de la Mothe, J. and Paquet, G. (eds) *Information, Innovation and Impacts*, Boston: Kluwer Academic Publishers.

Weingart, P. and Stehr, N. (2000). *Practicing Interdisciplinarity*. Toronto: University of Toronto Press.

Winter, S. (1987). "Knowledge and Competence as Strategic Assets", in Teece, D. (ed) *The Competitive Challenge: Strategy for Industrial Innovation and Renewal*, Cambridge MA.: Ballinger.

Wolfe, D.A. (1999). "The Role of Cooperative Industrial Policy in Canada and Ontario," in Braudo, R.J. and MacIntosh, J.G. (eds) *Competitive Industrial Development in the Age of Information*, London and New York: Routledge.

Woodcock, M. (1998). "Social Capital and Economic Development", *Theory and Society*, Vol. 2, No. 27.

INDEX

U

V

1. A. Phillips, A.P. Phillips and T.R. Phillips:
   Biz Jets. Technology and Market Structure in the Corporate Jet Aircraft Industry. 1994 ISBN 0-7923-2660-1
2. M.P. Feldman:
   The Geography of Innovation. 1994 ISBN 0-7923-2698-9
3. C. Antonelli:
   The Economics of Localized Technological Change and Industrial Dynamics. 1995 ISBN 0-7923-2910-4
4. G. Becher and S. Kuhlmann (eds.):
   Evaluation of Technology Policy Programmes in Germany. 1995 ISBN 0-7923-3115-X
5. B. Carlsson (ed.): *Technological Systems and Economic Performance: The Case of Factory Automation.* 1995 ISBN 0-7923-3512-0
6. G.E. Flueckiger: *Control, Information, and Technological Change.* 1995 ISBN 0-7923-3667-4
7. M. Teubal, D. Foray, M. Justman and E. Zuscovitch (eds.):
   Technological Infrastructure Policy. An International Perspective. 1996 ISBN 0-7923-3835-9
8. G. Eliasson:
   Firm Objectives, Controls and Organization. The Use of Information and the Transfer of Knowledge within the Firm. 1996 ISBN 0-7923-3870-7
9. X. Vence-Deza and J.S. Metcalfe (eds.):
   Wealth from Diversity. Innovation, Structural Change and Finance for Regional Development in Europe. 1996 ISBN 0-7923-4115-5
10. B. Carlsson (ed.):
    Technological Systems and Industrial Dynamics. 1997 ISBN 0-7923-9940-4
11. N.S. Vonortas:
    Cooperation in Research and Development. 1997 ISBN 0-7923-8042-8
12. P. Braunerhjelm and K. Ekholm (eds.):
    The Geography of Multinational Firms. 1998 ISBN 0-7923-8133-5
13. A. Varga:
    University Research and Regional Innovation: A Spatial Econometric Analysis of Academic Technology Transfers. 1998 ISBN 0-7923-8248-X
14. J. de la Mothe and G. Paquet (eds.):
    Local and Regional Systems of Innovation, 1998 ISBN 0-7923-8287-0
15. D. Gerbarg (ed.):
    The Economics, Technology and Content of Digital T V, 1999 ISBN 0-7923-8325-7
16. C. Edquist, L. Hommen and L. Tsipouri
    Public Technology Procurement and Innovation, 1999 ISBN 0-7923-8685-X
17. J. de la Mothe and G. Paquet (eds.):
    Information, Innovation and Impacts, 1999 ISBN 0-7923-8692-2

18. J. S. Metcalfe and I. Miles (eds.):
Innovation Systems in the Service Economy: Measurement and Case Study Analysis, 2000 ISBN 0-7923-7730-3
19. R. Svensson:
Success Strategies and Knowledge Transfer in Cross-Border Consulting Operations, 2000 ISBN 0-7923-7776-1
20. P. Braunerhjelm:
Knowledge Capital and the "New Economy": Firm Size, Performance and Network Production , 2000 ISBN 0-7923-7801-6
21. J. de la Mothe and J. Niosi (eds):
The Economic and Social Dynamics of Biotechnology, 2000 ISBN 0-7923-7922-5
22. B. Guilhon, (ed):
Technology and Markets for Knowledge: Knowledge Creation, Diffusion and Exchange within a Growing Economy, 2000 ISBN 0-7923-7202-6
23. M. Feldman and A. Link (eds.):
Innovation Policy in the Knowledge-Based Economy, 2001 ISBN 0-7923-7296-4
24. J. de la Mothe and D. Foray (eds.):
Knowledge Management in the Innovation Process 2001 ISBN 0-7923-7464-9

KLUWER ACADEMIC PUBLISHERS – BOSTON / DORDRECHT / LONDON